T0094228

MUSCLE

Also by Roy A. Meals, MD

Bones: Inside and Out

The Hand Owner's Manual

MUSCLE

The Gripping Story of Strength and Movement

Roy A. Meals, MD

W. W. NORTON & COMPANY
Celebrating a Century of Independent Publishing

Printed in the United States of America
First Edition

For information about special discounts for bulk purchases, please contact
W. W. Norton Special Sales at specialsales@wwnorton.com or 800-233-4830

Manufacturing by Lake Book Manufacturing
Book design by Lovedog Studio
Production manager: Lauren Abbate

ISBN 978-1-324-02144-5

W. W. Norton & Company, Inc., 500 Fifth Avenue, New York, N.Y. 10110
www.wwnorton.com

W. W. Norton & Company Ltd., 15 Carlisle Street, London W1D 3BS

1 2 3 4 5 6 7 8 9 0

To Susan, Clifton, Anderson, and Sydney

You inspire me to do my best and
to have fun in the process.

Contents

MUSCLE

Warm-Up

DID YOU JUST BLINK? WAS IT VOLUNTARY OR INVOLUNTARY? Either way, one set of tiny motors closed your eyelids, and another set opened them. As you read these words, the muscles in the iris of your eye are automatically moving to adjust the amount of light they let in, and those around your lens are focusing the words on your retina. These are examples of movement—one of life's essential features.

Sages as far back as Aristotle have tried—and failed—to come up with a universally acceptable definition of life. As a reasonable approximation, biology teachers resort to the mnemonic MRS GREN, which represents movement, reproduction, sensitivity, growth, respiration, excretion, and nutrition. These are the tightly interconnected index functions performed by all life forms. Among the seven attributes, movement plays double duty for animals, including humans. Not only is muscle movement critical to our internal workings of respiration, digestion, and reproduction, it also can transport us in search for the best air, food, and mate as well as remove us from danger. (This gives us a leg up on plants.) To appreciate our bodies, improve our health, and move artfully at all ages, an understanding of this rippling tissue and its myriad powers is paramount.

While it is possible for you to rest your eye muscles all night, another muscle never stops. Your heart has been contracting roughly 70–100 times a minute since you were a three-week-old embryo.

Cardiac muscle is amazingly durable and has the potential to sustain a human for over 100 years. Also working behind the scenes and out of our conscious control are the smooth muscles that cause us to blush, have goose bumps, and digest food. A third kind of muscle firmly attaches its ends to bones and can produce incredible feats of movement and strength. These skeletal muscles have enabled one man to complete 68 pull-ups in a minute, another to high-jump over 8 feet, and a woman to bicycle at a record speed of 184 miles per hour. Additionally, muscles allow for communication via facial expressions, hand gestures, and vocalization. In fact, your brain has no way of conveying its thoughts without contracting some muscles (unless, of course, you practice telepathy or subject yourself to an electroencephalogram or functional MRI). Whether keeping your blood flowing or propelling your legs when you go for a jog, muscle is the moving force behind our bodies' essential, day-to-day functions.

✦✦✦✦

Muscles stand apart not only for what they can do but also because—unlike so many of our bodies' internal elements—they can be observed. Only thinly draped with skin, muscles telegraph to an observer the person's overall health and vigor. The liver, kidneys, and other internal organs are just as vital as muscle, but unless they are way out of kilter, their states of health are not apparent from across the room. What's more, only muscle is amenable to spot training. With heavy lifting, you can develop bulging biceps, but heavy thinking won't enlarge your brain. (And although heavy drinking will enlarge your liver, it will be your whole liver, not just part of it, and detrimentally so.)

Regardless of your habits, are you happy with your weight? Strength? Physique? Blood pressure? Blood sugar level? Mental and physical endurance? Sleep pattern? Particularly if you are sedentary for most of the day, you likely have to answer "no" to at least one of these questions. Furthermore, do you want a long and active life? If so, healthy muscles are key.

Our muscle mass peaks at about age 27 and thereafter begins a long and inexorable decline, which, with good lifestyle choices, we can slow. Still, along the way, we may face a variety of maladies, including hypertension, myocardial infarction, gastric reflux, stress incontinence, or erectile dysfunction. All these problems stem from some muscle disorder, and a thorough understanding of how these derangements occur and can be treated guides informed lifestyle and treatment choices. Such understanding will have growing importance in future years as today's cutting-edge, emerging, and imagined technologies evolve and mature. These include perfecting artificial hearts, editing genes to cure muscular dystrophy, and advancing immunology understanding that will make pig-to-human heart transplants feasible.

Maladies aside, expected increases in life expectancy and leisure time will afford us the opportunity to enjoy our muscles wisely for longer. Yet, new health and wellness myths will arise, which in finality will not stand up to scientific scrutiny. There are also current misconceptions that die hard. My parents forbade me to go swimming for an hour after eating, supposedly to avoid muscle cramps that could lead to drowning. That warning, perhaps arising from a 1908 Boy Scout manual, has been thoroughly debunked. Both the American Academy of Pediatrics and the American Red Cross say it is fine to plunge right in after eating. (True, digestion draws circulation away from exercising muscles, which might cause a cramp, but not to a life-threatening extent.) More critically, diet supplements, exercise equipment, and gym memberships come with enticing claims for building muscle and losing fat, but how many of them are phony and which are scientifically based, life-maintaining, and life-enhancing advances that we should embrace? Answers to all these questions require an understanding of strength and movement.

Muscle: The Gripping Story of Strength and Movement is a guided tour through this force producer's myriad virtues and capabilities. Ranging between the disparate worlds of biology, art history, popular culture, and bodybuilding, as well as the frontiers of gene editing and stem cell research, you will come to understand and marvel at muscle's structure and function and be in position to understand new developments. We need, for example, to find a way for astronauts to maintain muscle mass during a zero-gravity trip to Mars. And is it true that the act of smiling can make you happier?

"Muscle" comes from the Latin word *mus*, meaning mouse, and was so named because it resembled a house mouse (Latin: *Mus musculus*) wiggling under the skin. These days, some men and women seek out not mice, but silicone implants to enhance their physiques. Nothing new here, because humans' infatuation with muscles spans millennia. Depicted in bronze and marble, ancient Grecian sculptures of Zeus, Atlas, Heracles, and fellow immortals are exceedingly buff, although it is unlikely that their sculptors had any direct knowledge of muscle anatomy. That changed in the Renaissance, and it shows. Michelangelo's muscular and lifelike *David* is a stunning tribute to the sculptor's genius and to Michelangelo's clandestine cadaver dissections. More recently, icons of pop culture, imagined and real, include Popeye, Superman, the Jolly Green Giant, Charles Atlas, and Steve Reeves—all with ripped physiques. All those who want to enhance their own musculature to emulate Marvel movie stars and those of us who just wish to maintain health and well-being without getting "swole" must first understand how our muscles work. Let's get moving.

Chapter 1

DISCOVERY AND DESCRIPTION

ANATOMY IS THE OLDEST MEDICAL SCIENCE, AND EVERY FIRST-year medical student continues the tradition via indoctrination in the gross anatomy lab. But gross anatomy is not repulsive. "Gross" in this context differentiates visible-to-the-unaided-human-eye anatomy from microscopic study. A quick look across time will be revealing.

Sushruta, India's equivalent to Hippocrates but about 400 years earlier, wrote that every medical student should dissect a human body. The prevailing attitude, however, was that bodies of the recently deceased were sacrosanct and should be touched only in preparation for burial. Sushruta advised a workaround for inquisitive medical students: place the corpse in a wicker basket and submerge it in a slowly moving stream. Some days later, briskly whisk-broom the decomposing body without touching it to reveal its anatomical secrets. That undoubtedly *was* gross, and the process did not reveal and preserve any useful knowledge of muscle anatomy.

Neither did the practices in ancient China and Egypt. Confucius dominated intellectual life in China for centuries, and reasoning and assumption prevailed over observation, which would have meant dissection. In ancient Egypt, perhaps 70 million bodies were eviscerated and embalmed and certainly could have provided ample material for methodical dissection and advanced understanding. However, a special class of priests, rather than doctors, prepared the dead for burial and removed the internal organs piecemeal through small openings, again missing an awesome opportunity to study anatomy and pathology.

Galen's Pervasive Influence

It was the same in ancient Greece as it was in China and Egypt. Knowledge was based on logic and speculation. Reasoning superseded observation. The advent of observational anatomy took place across the Mediterranean in Alexandria, where in about 250 BCE Herophilus and Erasistratus likely performed the first systematic human dissection. Observational science, however, did not catch on in the Roman Empire, where the Greek tenets regarding knowledge and learning prevailed.

Then, about 150 CE, along came Galen, a Greek doctor living in Rome. In addition to being physician to emperors and surgeon to gladiators, he was a prolific medical writer, the most influential of all time. Unfortunately, regarding human anatomy, he was often wrong. This stemmed from his minimal firsthand observations of human anatomy, which were limited to gladiator wounds and the occasional corpse that washed up onshore. Galen compensated for this by dissecting monkeys, pigs, and bears. He assumed, erroneously, that human anatomy was identical to that seen in his dissections.

More than a thousand years later, physicians and anatomists still held Galen's writings in the highest esteem, and when they encountered a discrepancy between Galen's "truth" and their own lying eyes, they sided with Galen. For instance, they rationalized that the anatomy they were observing had apparently changed since Galen's time and that the curved thigh bone described by Galen had straightened over the centuries by the habit of wearing "cylindrical nether garments." Medieval anatomists also reasoned, when they found fewer segments constituting the breastbone than Galen had recorded, that the robust chests of ancient heroes might well have had more bones than current-day degenerates could claim.

It should not be surprising then that anatomical drawings of the

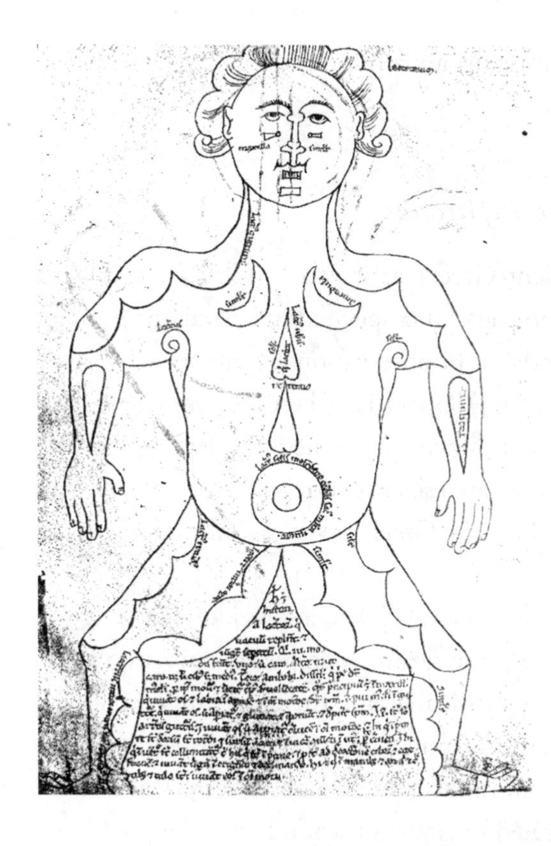

This thirteenth-century depiction of muscles fairly accurately represents the deltoids and thigh muscles but completely misses the "six-pack" abdominal muscles, unless maybe the artist was being ironic by representing them as a bagel.

time were crude sketches at best if not entire fantasies. To the credit of the pre-Renaissance anatomists and physicians, however, there was no great need to understand anatomy, especially musculoskeletal anatomy, because bloodletting and amputations were the only limb surgeries performed.

Anatomy and Art

Anatomy, like so many fields of study, changed with the coming of the Renaissance. Artists and anatomists became more interested in understanding and depicting the human body, including its internal structures. In the 1470s, Antonio del Pollaiuolo, who is known along with his artist brother to have dissected corpses, painted the *Battle of the Nudes*. It reveals a remarkable understanding of straining muscles and has been described as "a rhetoric of anatomical display that would last

Renaissance artists began taking an interest in portraying human musculature. Top: Battle of the Nudes *by Antonio del Pollaiuolo, c. 1470. Lower left:* Saint Sebastian at the Column. *Albrecht Dürer, 1500. Lower right:* Saint Sebastian Tied to a Tree. *Albrecht Dürer, c. 1501.*

for centuries." Thirty years later, the German painter Albrecht Dürer, among others, reinforced the melding of art and anatomy when he began portraying realistic nude bodies, under the guise of telling Christian stories to avoid religious sanctions.

The intertwining of the two disciplines came to full flower during the Renaissance and especially through the contributions of Andreas Vesalius (1514–1564). He was a quick study. On the day he graduated from medical school, not quite 23 years old, he accepted the offer to become the chair of anatomy and surgery at the University of Padua. Six years later he published the seminal *De Humani Corporis Fabrica Libri Septem* (*On the Structure of the Human Body in Seven Books*). After 1,400 years, finally somebody had the confidence and gall to refute Galen! Inspection and measurement finally superseded reasoning and speculation, and Vesalius's book heralded the beginning of modern observational science and research. No longer would reasoning usurp observation.

Most of the observations, however, were clandestine because the Catholic Church strongly prohibited human dissection. The exception was executed criminals. Medieval medical schools were allowed one dissection per year, which was performed publicly, apparently to deter would-be criminals. Even almost 300 years later, the cursed sentence for an assassin remained, "That you be taken from hence . . . to a place of execution, where you shall be hanged by the neck until you be dead; your body to be dissected and anatomised."

Once Vesalius had dissected a cadaver, he posed it in a lifelike position. In *Fabrica*, sequential plates depict muscles folded back layer by layer to reveal their relationships to one another. Vesalius identified muscles mostly by numbers—for instance, "muscle raising the arm, the second of those that move it" and "first muscle moving the foot." Later, others named these the deltoid and gastroc-

Andreas Vesalius depicted his anatomical specimens in lifelike poses, for the most part numbered the muscles, and folded some of them back to expose those located underneath. The background landscape makes a panorama when plates from Fabrica *are placed side by side.*

nemius (calf muscle), respectively. Vesalius did, however, give names to some muscles that persist today, including rectus abdominis (the muscles used for sit-ups) as well as masseter and temporalis (both for chewing).

Vesalius was the premier Renaissance anatomist and depicted his dissections with an artist's flair. Vesalius and Michelangelo (1475–1564) died in the same year. Did they know each other and share ideas? I don't know, but it is obvious that Michelangelo, in both paint and marble, portrayed his subjects with an anatomist's accuracy. Although he took inspiration from the ancient Greek and Roman statuary that still existed in Italy during his time, he initially felt constrained. He lamented the Catholic Church's strict edict against human dissection. Michelangelo knew that he could never portray the human body realistically without knowing its interior structure. Fortunately for the world's cultural benefit, he did dissect, on the sly. (Read the memorable details in Irving Stone's biography of Michelangelo, *The Agony and the Ecstasy*.)

The two disciplines of art and anatomy truly melded under the pen of Leonardo da Vinci (1452–1519). Among his seemingly myriad pursuits, perhaps human anatomy piqued his interest more than any other subject. He said that he dissected more than thirty cadavers over his lifetime. His drawings, sketches, and accompanying descriptions show his deep understanding of the human musculoskeletal system *in motion*, which was an entirely novel approach.

His studies, however, were not published until after his death, and some were lost until the nineteenth century, so his influence on later Renaissance anatomists and artists was blunted. On discovery, however, one further aspect of Leonardo's genius became evident. He understood the muscles' attachments to bone, particularly around the shoulder, and the lever actions so created. He appreciated that a muscle's function varied according to the starting position of the joint(s) it controlled—the anatomy of movement. Leonardo also dabbled with cross-sectional musculoskeletal anatomy, a mode of representation that would not become clinically useful until the invention of computed tomography and magnetic resonance imaging nearly 500 years later.

Renaissance artists not only developed and capitalized on the techniques of perspective and shading, they moved away from portraying just religious figures. If not performing dissections on their own, the artists enhanced their understanding of human anatomy by attending the once-a-year public dissections of executed criminals allowed by the Church.

As happened repeatedly in Galen's writings, errors in anatomical descriptions and depictions continued to occur. A famous one appears in Rembrandt's *Anatomy Lesson of Dr. Tulp*, painted in 1632. The forearm muscles responsible for closing the fist are seen to be attached on the outside of the elbow, whereas in fact those muscles attach on the side of the elbow closest to the trunk when the arm is at the side. True, this detail does not diminish the painting's impact (except maybe for hand surgeons), but it is a curiosity that possibly arose from Rem-

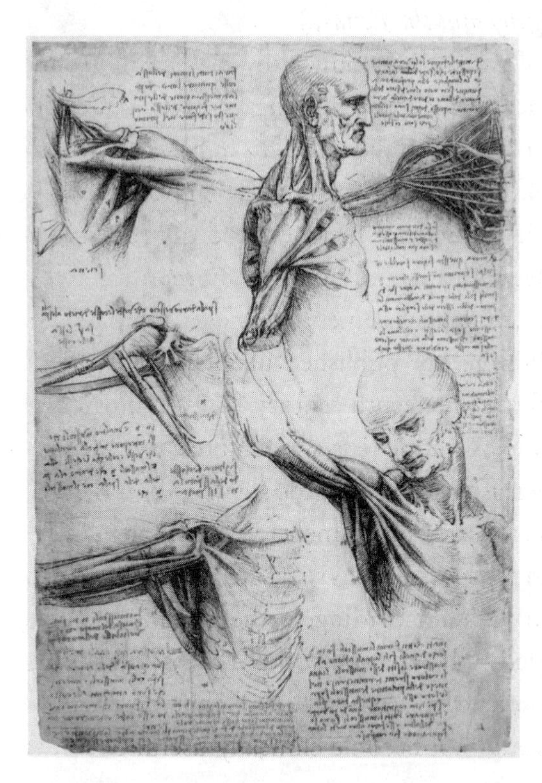

Leonardo da Vinci's drawing and notes indicate his deep and novel understanding of the anatomy of motion, which is particularly relevant for the freely moving shoulder. When the joint is in one position, for example, with the arm at the side, a muscle may have one function yet have an entirely different effect when the shoulder is holding the arm overhead.

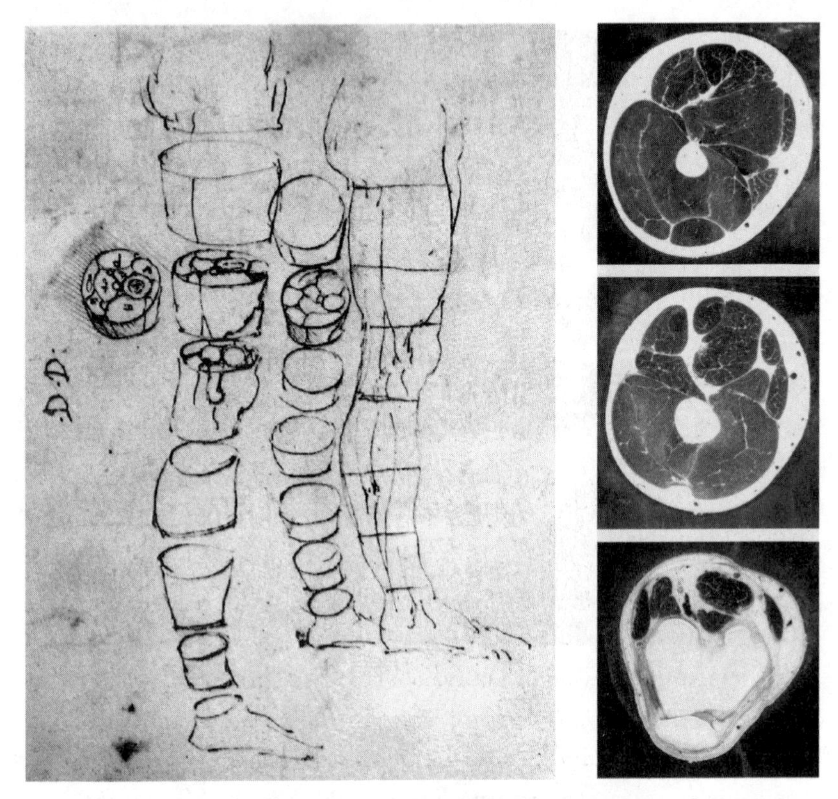

Left: Leonardo da Vinci appreciated that representation of cross-sectional anatomy could display the undisturbed relationships among the muscles, bones, vessels, and nerves. Right: Top to bottom, cross sections from the Visible Human Male Project through mid thigh, distal thigh, knee.

brandt first sketching the anatomy of a right forearm and then not correctly reversing the muscles' orientation in the final painting.

Depiction in Three Dimensions

Despite the great advances encountered during the Renaissance in understanding anatomy and depicting the findings artistically, problems persisted in trying to depict real-world three-dimensional anatomy in two-dimensional paintings and drawings. To an extent, this was circumvented in the eighteenth and nineteenth centuries when

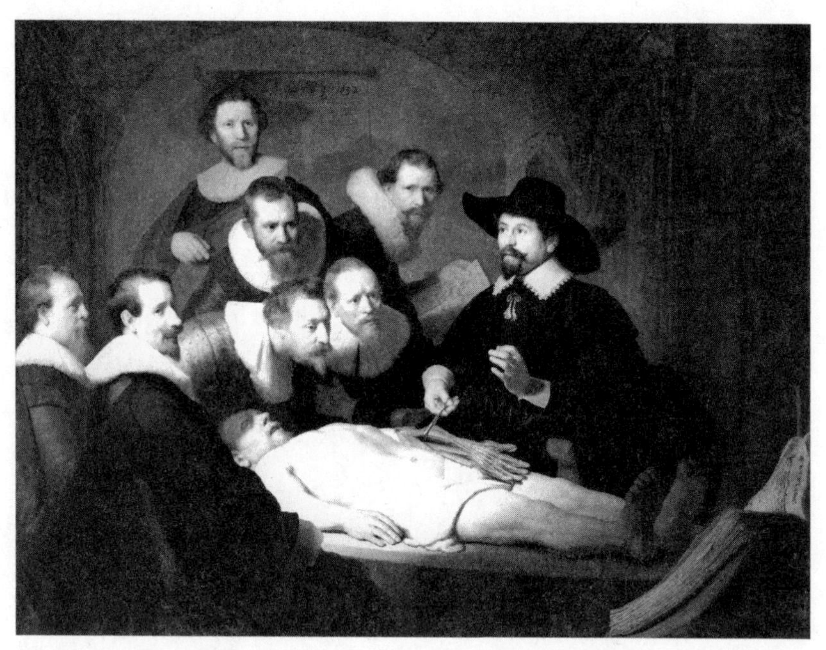

The Anatomy Lesson of Dr. Nicolaes Tulp *by Rembrandt van Rijn, 1632. Dr. Tulp is demonstrating the forearm anatomy to a group of Amsterdam physicians, some of whom paid commissions to be included in the painting. The open book in the lower right may well be Vesalius's* De Humani Corporis Fabrica.

European medical students used full-scale wax and papier-mâché models to study human anatomy. This method did not sustain interest, but another one did and today continues to grow in importance.

Russian Nikolai Pirogov (1810–1881) likely did not know of Leonardo's cross-sectional sketches when he embarked on making his cross-sectional anatomy atlas, *An Illustrated Topographic Anatomy of Saw Cuts Made in Three Dimensions across the Frozen Human Body,* which he published in 1855. Nine years before, while visiting a butcher's shop in the dead of winter, Pirogov noted that frozen pig carcasses sawed straight across demonstrated the precise positions of anatomical elements with respect to one another, whereas conventional dissections necessarily disturbed the locations of superficial structures to view deeper structures and thereby destroyed their special relation-

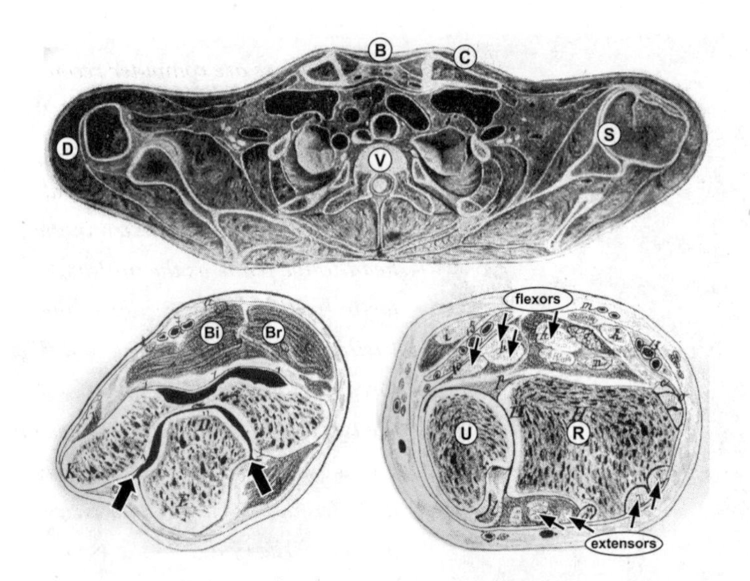

These cross-section drawings are from Pirogov's 1855 anatomy atlas.
Above: This section shows the shoulder joints (S), deltoid muscle (D), vertebra (V), and breast- and collarbones (B, C).
Left: A section through the elbow joint shows the biceps (Bi) and brachialis (Br) muscles lying anterior to the skeleton.
Right: In the forearm near the wrist, the ulna (U) and radius (R) separate the tendons that close the fingers (flexors) from those that open them (extensors).

ships. Pirogov realized that he could use the cold Russian winters to freeze human cadavers "to the density of the thickest wood" and then saw them into slices sometimes as thin as one-sixteenth of an inch, thaw them, and accurately draw the anatomical structures with their spatial relationships to one another fully preserved. Like Leonardo, Pirogov recognized the medical value of such cross sections over 100 years before computed tomography and magnetic resonance imaging came along, whose raw images are entirely cross sections.

In the 1990s, the National Library of Medicine extended the work of da Vinci and Pirogov by freezing a man's body, and later a woman's body, in huge blocks of ice. The anatomists then ground away the blocks 1 mm (1/25 of an inch) at a time from head to toe

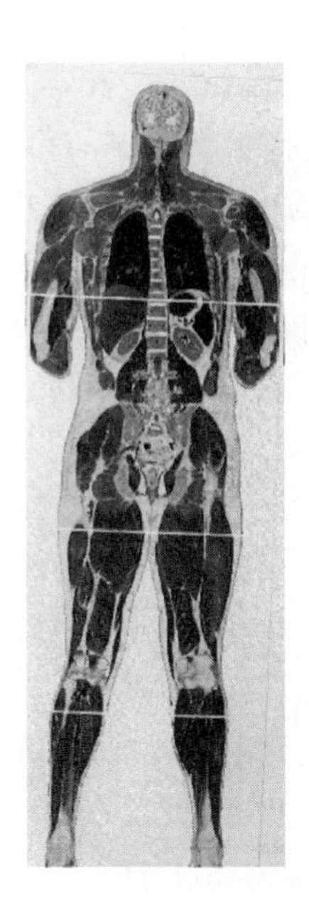

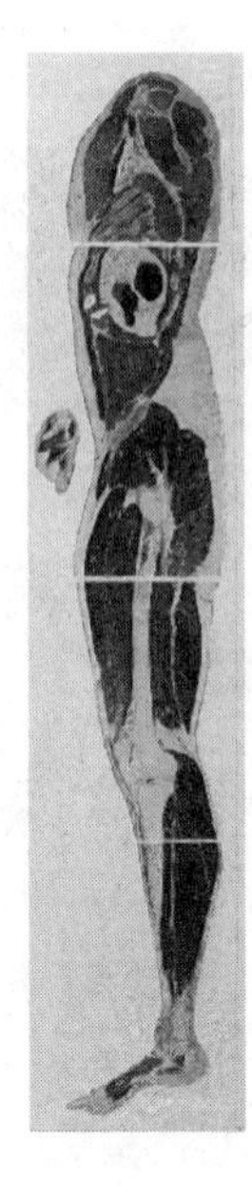

These images are computer reconstructions in the front-to-back plane (left) and side-to-side plane (right) obtained from the original photographs that were taken in the head-to-toe plane as the millimeter-by-millimeter dissection of the Visible Human Male proceeded. The forearms and hands are missing in the left image because they rested on the cadaver's abdomen and were not included in this cut. The hands can be seen in the image on the right.

for the man and at three times closer intervals for the woman. At each interval, they momentarily thawed the block's surface and photographed the anatomy. For the man, who had been slightly over six feet tall in life, this resulted in nearly 1,900 images, which make for a stunning head-to-toe tour of human anatomy when viewed in the time-lapse cine. The anatomists also manipulated the digital representations in the computer to produce accurate images in both the front-to-back and side-to-side planes. The body donor for the Visible Human Male was a murderer in life, so, apropos of medieval and Renaissance times, his body was thoroughly "anatomised."

Other donors have had their bodies "plastinated" and then taken on world tours. Plastination had its start in the 1970s. A German anatomist, Gunther von Hagens, invented the technique for removing water and fat from fresh cadavers by vacuum and then replacing these materials with curable polymers, which left the skinless tissues firm to the touch and odor free. Von Hagens then molded the bodies into lifelike poses, some mimicking Vesalius's figures, others in full action poses, including an equestrienne (with her horse, also plastinated, in full stride) and an athlete at the moment of javelin release.

Whether or not you have previously considered anatomy "gross," these traveling exhibits in natural history and science museums celebrate anatomy in a way that nonmedical people seem to find interesting. When I have attended, other visitors from age six on up have expressed nothing but awe and respect for the incredible displays of skeletal musculature seemingly at work. Anyone who attends one of these exhibits is likely to be impressed and enlightened.

Muscles' Monikers

While interest in and knowledge of anatomy increased steadily during the Renaissance, differentiating and naming the newly observed muscles proceeded with fits and starts. Vesalius tended to number them. In addition to naming two jaw muscles and the six-pack, he also named an arm muscle—the anterior cubitum flectentium musculus—though it seems, in this case, a nice succinct number would have been more user-friendly. Had all the muscles retained numbers, however, it could get cumbersome if somebody asked you to flex your number 489 and you couldn't remember which one it was, out of the roughly 650 that humans have.

Fortunately, anatomists after Vesalius pitched in and gave the muscles descriptive names and renamed the anterior cubitum flectentium musculus the biceps. That's part of the good news. The bad news is that because Latin was the language of science at the time, some of the names may seem foreign to those of us whose Latin skills are languishing. The other part of the good news is that with a uniform Latin terminology, anatomists and health-care providers around the world could and continue to communicate clearly with one another; and with a bit of linguistic dissection, the names are not all that foreign.

Some of the original names were outright poetic. Consider, for instance, the contributions of Jan Jesenius (1566–1621), a Bohemian physician, politician, and philosopher. He named the muscles controlling the eyeball's movement *amatorius* (muscle of lovers), *superbus* (proud muscle), *bibitorius* (muscle of drinkers), *indignato-*

rius (muscle of anger), and *humilis* (muscle of lowliness). Twenty years later Jesenius was executed, but it was his political allegiance rather than his muscle naming that got him in trouble. It is too bad that in 1895 anatomists standardized the nomenclature and dully renamed the eye muscles according to their location (superior, inferior, medial, lateral) and alignment (rectus [straight] and oblique). In fact, contraction of the medial rectus muscles would make one cross-eyed, so maybe the Terminology Committee should have left them as the muscle of drinkers (*bibitorius*).

Most names are straightforward and need only a smattering of Latin to understand. For instance, some muscles received names according to their location, such as the subclavius (under the clavicle) and the intercostales externi (external layer, between the ribs). Others are named by the number of their parts: *Bi-* means two, and the biceps has two origins, one from the shoulder blade, one from the upper arm bone. The triceps has three origins, and the quadriceps has . . . well, guess.

Length determines the name for some muscles. *Thumb* in Latin is *pollux*, and it has two muscles that fold (flex) the thumb across the palm—the flexor pollicis longus and flexor pollicis brevis. Size matters too. I am guessing that you are presently sitting on your gluteus maximus muscles. (*Gluteus* comes from Greek *gloutos*, buttock.) Between the gluteus maximus and the pelvis are the gluteus medius and gluteus minimus muscles. And smack against the back of the hip joint is a matched pair of small muscles, the superior and inferior gemelli—twins.

In the abdominal wall, alignment is everything. The rectus (straight) abdominis (the highly valued "six-pack") runs longitudinally, while the obliquus externus abdominis and transversus abdominis run their separate routes.

Other muscles resemble geometric shapes and are so named. There are three "quadratus" (square) muscles. One is in the foot, one is deep in the forearm, and the other crosses the hip joint. The rhomboid major and minor are parallelogram-shaped muscles that attach on the thoracic spine and the shoulder blade; and the deltoid, crossing over the top of the shoulder, is the shape of the Greek letter Δ. The serratus anterior has a jagged origin from multiple ribs on the front of the chest, and the gracilis is indeed slender or gracile—a long, thin muscle on the inner thigh.

Several muscles' names describe their action. The cremaster (a muscle that lifts the testicle) derives its name from the Greek verb for "I hang," and the levator scapulae raises the shoulder blade.

A few muscles' names simply identify their origins and insertions. For instance, the sternocleidomastoid is the strappy muscle on the side of the neck that turns your head to the side. One end attaches on the breastbone (sterno) and collarbone (clavicle, cleido) and the other end fastens to the mastoid process of the skull, which is palpable just behind the earlobe.

Some muscles received names of objects they resemble. Piriformis, a hip muscle, is pear-shaped. The deep calf muscle, the soleus, is sandal-shaped. Overlying it is the bulgy gastrocnemius, literally the belly of the leg. In each palm and sole are four worm-shaped muscles, lumbricales manus and lumbricales pedis, respectively. The Latin name for earthworm is *Lumbricus.*

My favorite muscle name is sartorius, which applies to the longest muscle in the body. It starts high on the pelvic rim, crosses the front of the thigh, and finishes on the inside of the leg just below the knee. Contracting the sartorius on both sides causes the hips to flex, the thighs to rotate outward, and the knees to flex, resulting in the owner ending up in a cross-leg sitting position. This is the position that tailors traditionally assumed when working on garments in their laps. *Sartor* in Latin means tailor.

By the eighteenth century, anatomists' interest in gross anatomy had waned. They had identified and named nearly everything, and so they turned their attention to fetal development and the anatomy of disease. During the last half of the nineteenth century, muscle attracted the interest of biochemists, who slowly have unraveled how muscle creates movement.

Chapter 2

MOLECULAR MAGIC

TODAY A GRASP OF MUSCLE'S CAPACITY FOR STRENGTH AND movement requires an understanding of its microscopic structure and chemistry. It is heady stuff and has fueled entire careers for many scientists and has—so far—yielded three Nobel prizes. The description of how muscle works on a molecular level is by far the most technically challenging material in the book. My explanations of the current consensus understanding are as non-technical as I can make them. Although muscle physiologists may grimace at my simplifications and analogies, I feel that helping non-specialists grasp the basic concepts of how muscle works is worth the possibility of incurring a specialist's disdain. In fact, the way the two principal molecules in muscle work together underpins everything about muscle, from bodybuilding and "hitting the wall" to heart attacks and rigor mortis. Mastering the basics will be worth the effort. Be strong.

Muscle, an astounding molecular motor, consists of two protein filaments, one that advances stepwise along the other and then retreats. By converting chemical energy into physical force, these reciprocating protein filaments have been generating animal movement for 600 million years, beginning with jellyfish and extending to earthworms, snails, fish (where muscle may comprise 60 percent of their body weight), and us. Their repeated interactions propel fleas high in the air and transport track stars a mile in less than four minutes.

One of the proteins, actin, is the most common protein in most plant and animal cells, which makes it the second most common protein on earth. (The first is a photosynthesis-inducing enzyme in green plants.)

Actin consists of roughly 375 amino acids in a chain, and its amino acid sequence is 94 percent identical in organisms as widely varied as yeast and yaks. Considering that yeast cells have divided once every two hours for a billion years, that adds up to gazillions of cell divisions and opportunities for mutations to occur, yet from early on actin already had it right. In other words, the inevitable genetic mutations that caused later changes in actin were principally unfavorable and therefore not passed on. In molecular biology parlance, the structure of actin has been *conserved.*

Myosin is actin's partner in movement. Of course, it mutated too, but at least 14 variations of myosin found separate purposes and are at work today. Only one, however, matches up with actin in muscle cells.

Rowing to Iceland

I hope this analogy will help you understand how actin and myosin work together. Imagine a short double-ended pen. Along the shaft of the pen where you would grip it are closely spaced "oars" sticking out, like those on a Viking ship. The "rowers" sitting toward one end of the pen are facing those sitting toward the other end. If the pen were in water, the opposing teams of rowers would be pulling against each other, swirling the water, but not moving the boat.

For the pen, imagine that it has loosely applied caps, which just barely cover each end of the pen. Inside each cap, multiple pockmarks line the surface for half of the cap's depth. When the imaginary helmsman shouts "stroke," the rowers closest to each end of the pen catch the tips of their oars in the pockmark just inside the open ends of the caps. This stroke pulls the caps ever so slightly onto the pen. The helmsman commands "recover, stroke." The pre-

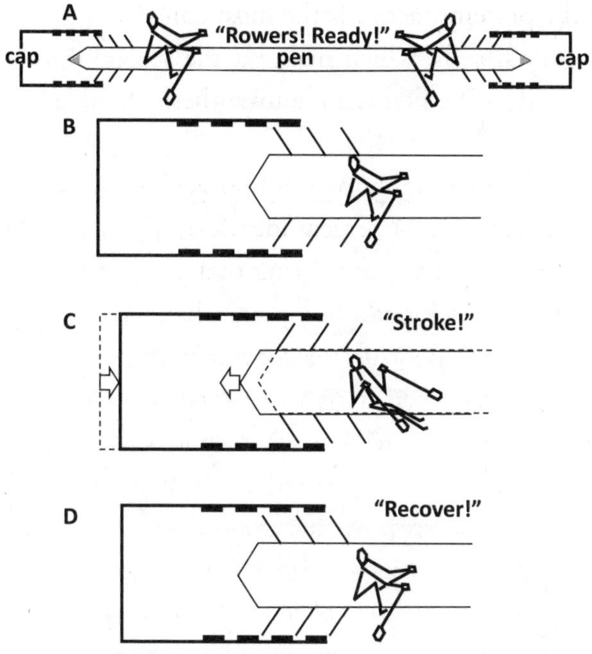

A. Caps on a double-ended pen with "rowers" facing each other.
B. Enlarged detail of A. One pair of oars engages in the first set of pockmarks in the cap. The oars at the opposite end of the pen (not depicted) are mirroring the illustrated engagement.
C. With a stroke of the oars, the cap and pen move closer together.
D. On the recovery stroke, the oars advance into the next pockmark on the cap. Now two pairs of oars have purchase on the cap.

viously engaged oars are released from their pockmarks, return to their original "cocked" positions, and then reengage in the next set of pockmarks along the line. At the same time, the just-vacated pockmarks provide "oar holds" for the next set of oars, and the cap-pen-cap assembly shortens a bit more. "Stroke, recover, stroke!" The oars repeatedly engage-release-engage the series of pockmarks, and the caps move progressively toward each other. Finally, despite the helmsman's orders, there is no more movement—the pen has moved into each cap beyond the pockmarked areas, so there are no more oar holds available. "Okay, rest." All the rowers stow their oars, which completely disengages the pen from the caps. The caps

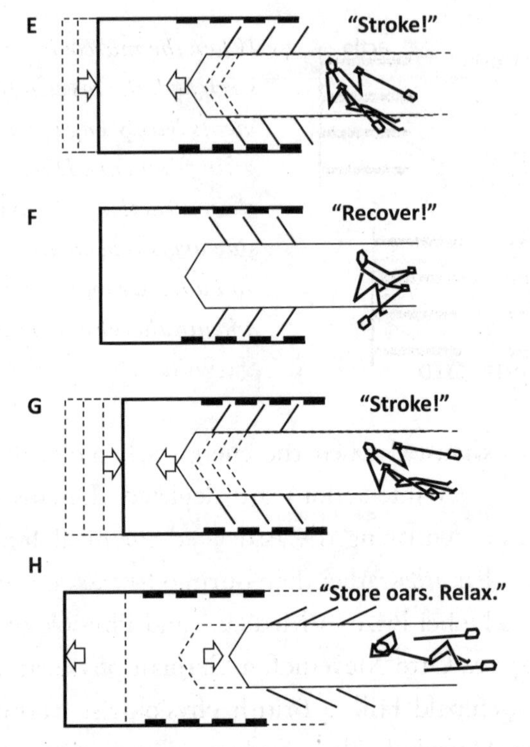

E, F, G. With each successive stroke, the cap and pen move progressively closer together.
H. On relaxation, the oars disengage from the cap, allowing the cap and pen to return to their original positions.

slide back to their original positions and again only loosely cover the pen's tips.

You have witnessed the essence of a muscle contraction, except that there are obviously no pens and caps in our bodies. The pen represents a thick protein filament (myosin), and the caps represent thinner protein filaments (actin). When stimulated by one of a variety of energy-releasing molecules, the myosin filament converts chemical energy into movement by changing the configuration of its short side branches and "rows" its way along the actin filaments. The unit shortens.

This chemical-to-mechanical reaction generates heat and carbon dioxide. It should be no wonder then that intense muscle activity—running, for instance—will leave a person flushed and breathless

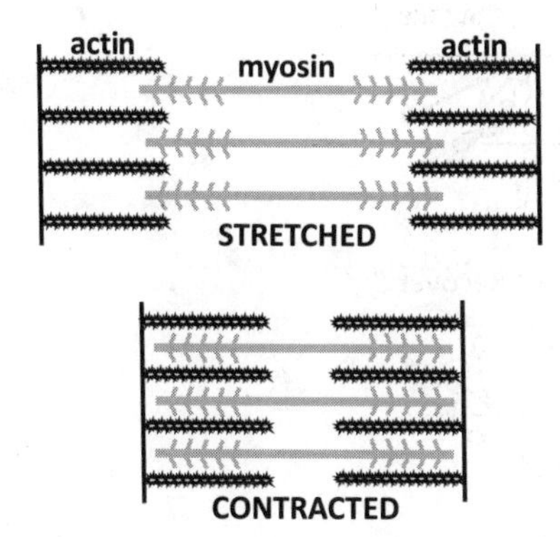

When the muscle is stretched, the myosin filaments barely engage with the actin filaments. During muscle contraction, the myosin's side arms repeatedly engage and advance on the actin filaments, thereby shortening the unit.

and eventually exhausted when the energy-rich molecules powering the actin/myosin interactions are depleted. Discovering this phenomenon and identifying the associated chemical conversions, some of which occur *after* rather than during the exercise, led to two men sharing the Nobel Prize in Medicine and Physiology awarded in 1922. They were Otto Meyerhof, a German physician and biochemist, and Archibald Hill, a British physiologist working independently from Meyerhof. Their findings significantly overlapped and confirmed each other's. As many researchers in biology and medicine have done in the past, Hill used himself as a subject. He ran sprints every morning for three hours to study the way his muscles consumed energy and oxygen. He learned that increased oxygen consumption during rest restored his energy stores.

The conversion of chemical energy to movement occurs on the molecular level, and we know that molecules are minuscule. So isn't this actin/myosin unit too small to power a breaching whale or even a blinking eye? More specifically, how big is this sublime and ubiquitous motor unit? Lined up end to end, 120 of the actin/myosin units could reach across a grain of table salt. That is the same whether the units are from a gnat or a gnu. Upon receiving the message to contract, each unit can shorten by 10–25 percent. That amount of change on a grain of salt would go unnoticed. How then can myosin filaments stepping along actin filaments power jellyfish and weight lifters? Teamwork! When millions of actin/myosin units connected end to end contract simultaneously, the entire muscle shortens by as much as 25 percent. By combining their contractions, the actin/myosin units can complete a biceps curl. When millions of these units are arranged principally side to side rather than end to end, the muscle will not contract by a great distance but will do so with great force, powering sit-ups, for instance.

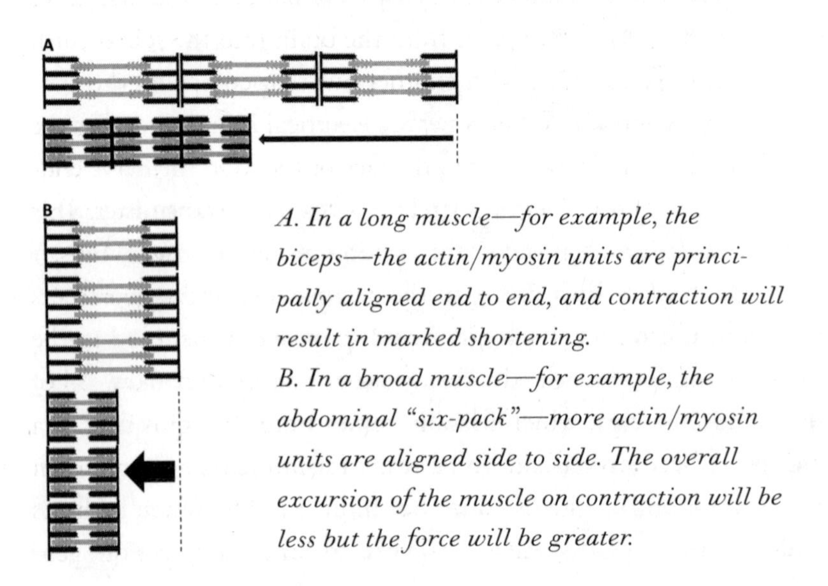

A. In a long muscle—for example, the biceps—the actin/myosin units are principally aligned end to end, and contraction will result in marked shortening.
B. In a broad muscle—for example, the abdominal "six-pack"—more actin/myosin units are aligned side to side. The overall excursion of the muscle on contraction will be less but the force will be greater.

If you listen carefully, you can hear your actin and myosin at work. Grimace forcefully by squinting and trying to bring the corners of your mouth to your ears, as if you just tasted the yuckiest thing ever. Do you hear a faint roar like a distant waterfall? It is the myosin molecules advancing on the actin filaments in your facial muscles.

Contraction Control

What controls these miracle molecular motors? In short, electrical impulses. Where they come from and how they get to the actin/myosin unit is varied and complex. For skeletal muscles, with which we are most familiar, the electrical message starts with an idea in the motor-cortex portion of the brain. This area is situated just inside the skull above the ear and sends commands to nerve fibers that connect to skeletal muscles. For muscles in the face, the connection is direct. A single nerve fiber, exiting the cranium through one of several small openings, carries the electrical impulse the entire way from brain to muscle. For muscles in the trunk and limbs, the connection requires an intermediate synapse, or handoff. The first nerve transmits the electrical impulse from the brain into the spinal cord. There the nerve connects with another one that completes the connection to the muscle. When a nerve's electrical impulse reaches the surface of the muscle, whether in the face or the foot, the nerve endings release a chemical appropriately called a neurotransmitter. This sets off a complex chemical chain reaction inside the muscle cells.

In the Viking ship analogy, the helmsman's shout represents the neurotransmitter. Each command to stroke floats away in the wind and must be repeated to bring about successive strokes. All of this requires energy, which initially comes from the equivalent of a molecular-sized power bar. To be ready for immediate action, each rower sleeps with a bite of power bar in his mouth. When he hears "Stroke," he gulps this energy packet and jumps into action, generating both mechanical force and heat without consuming much oxygen. Each stroke requires another bite, and the rowers entirely consume the power bars in one to three minutes, which is enough time to speed the Viking vessel across a wide river.

Now the helmsman decides to venture into the bay, but the rowers are out of power bars. From the hold, the helmsman distributes a barrel of sweet, sticky substance called glycogen. It quickly converts to sugar, and the crew members gain the necessary energy to get back to work. When the glycogen is gone, the exhausted crew gasp in Old Norse, "We've hit the wall." Now they must slow down or stop until the baker brings more power bars from the galley or until someone can find another barrel of glycogen. Mercifully, the helmsman announces, "Just slow and steady now. Maintain this pace because it must take us all the way to Iceland."

Throughout the early part of the voyage, ship attendants circulate among the rowers offering tanks of oxygen and trays heaped with cookies, candy bars, and bottles of sports drink. Although the power bar did not require any oxygen to transform chemical into mechanical energy, carbs and fat do; but over the long haul, carbs and fat provide a far more sustainable and efficient means of power conversion. They just don't get started as quickly. With plenty of glucose, fat, and oxygen molecules circulating, the rowers catch their second wind and power the craft far into the North Atlantic.

Before arrival in what will eventually become Reykjavik, however, they encounter unanticipated head winds. Carbs and fat are now in short supply. The only energy source left to the crew members is their own protein, including their actin and myosin, which of course have been responsible for their progress. Let's hope that these now skinny voyagers make port before they waste away entirely.

To set things straight, muscles do not really consume power bars. Rather, the muscle's immediately available energy source is a molecule called ATP—adenosine triphosphate. It's not particularly important to remember the name of this energy-laden chemical. Just note that the "tri" indicates that ATP includes three phosphate molecules. When a chemical chain reaction knocks a phosphate off the ATP molecule, it releases energy, which causes the side arm on the myosin filament to change shape and advance a step along the actin.

The good news about ATP is that it is stored adjacent to the myosin, right where it is needed, and that this energy-releasing reaction does not require oxygen. Therefore, the muscle's contractions are immediate, forceful, and rapid—as many as 70 contractions per second—which accounts for the faint roar you hear when you grimace forcefully. The bad news is that the supply of ATP runs out after just a couple of minutes. That is plenty of time for a startled duck to rise off the lake or a sprinter to run 400 meters; but after that, muscles fatigue rapidly and can continue by only one of two means.

The first is by resting for a few minutes and allowing a new supply of ATP to form, which permits sprinters to run in more than one heat or race on the same day. Alternatively, the muscle can continue working without rest but more slowly and after switching to another power supply. That energy source is glycogen, which is a gooey assemblage of glucose molecules and fat. Rested muscle cells have a supply of each, but they can convert neither glycogen nor fat into energy as quickly as they can using ATP. Also, glycogen needs to react with oxygen to release energy and change the myosin's shape, so this force-producing process is slower (no more than 30 contractions per second) and requires an abundance of blood vessels to supply the needed oxygen. More good news—this type of muscle contraction can go on for hours, provided that the blood can deliver enough oxygen and that the glycogen stores are sufficient. This is how marathoners keep going, but their recovery to pre-race condition is measured in weeks and months compared to a sprinter needing only hours or a few days to rebound.

When glycogen in the muscle is depleted, glycogen in the liver comes to the rescue. Once that is gone, the tank is empty. Endurance athletes describe this queasy, weak feeling as "hitting the wall," and it means they will need to rest and resupply. In the absence of a new supply over days and weeks, the body, now in starvation mode, turns first to fat. When that is gone, the body begins breaking down its proteins, including the actin and myosin in muscle. That is why starvation results in being nothing more than "skin and bones."

To stay on a somber note for a moment, actin and myosin continue to interact after death and account for the phenomenon of rigor mortis, which is the reason why a dead body is sometimes described as a "stiff." During life, energy is required not only for the myosin to take each step along the actin but also for the myosin to disengage and reset itself for the next advancement. At death the muscle's oxygen supply is cut off. Chemical energy stored locally, however, can power the myosin's catch-release actions without oxygen for the next several hours, longer in cold environments. When that energy source is depleted, the myosin cannot disengage from the actin, and muscle stiffness ensues. Depending on the temperature, rigor mortis peaks at about twelve hours after death and dissipates after forty-eight hours when the actin and myosin, along with other proteins, begin to disintegrate. Understanding this phenomenon and its chronology helps medical examiners and homicide detectives determine time of death.

However, that task might be tough if a mystery case involved a Greenland shark, which has a life span of 250 to 500 years, all of it in icy waters. Henry William Dewhurst wrote, in 1834, "When hoisted upon deck, it [a Greenland shark] beats so violently with its tail, that it is dangerous to be near it, and the seamen generally dispatch it, without much loss of time. The pieces that are cut off exhibit a contraction of their muscular fibres for some time after life is extinct. It is, therefore, extremely difficult to kill, and unsafe to trust the hand within its mouth, even when the head is cut off. And, if we are to believe Crantz, this motion is to be observed three days after, if the part is trod on or struck." Apparently, a postmortem mechanical stimulus creates an electrical discharge that activates cryopreserved and still functioning actin/myosin units.

Muscles *during* life, however, are far more interesting. Let's become familiar with several concepts that come up repeatedly.

Lifting versus Holding

Consider how you stand still while holding a suitcase. You probably never thought that was a big deal. In fact, to keep from falling over, you "recruit" several muscles to resist gravity, yet the muscles are neither shortening nor lengthening, and the joints in your arm, back, and legs do not move. These are isometric (same-distance) muscle activations. Contrast that to isotonic (same-tension) recruitment, which results in the activated muscles changing length, moving joints, and thereby raising or lowering the suitcase.

Furthermore, isotonic contractions come in two forms, concentric and eccentric. Concentric is what happens when you *lift* the suitcase—the muscles in your arm and shoulder shorten, the joints move accordingly, and the suitcase rises. An eccentric contraction

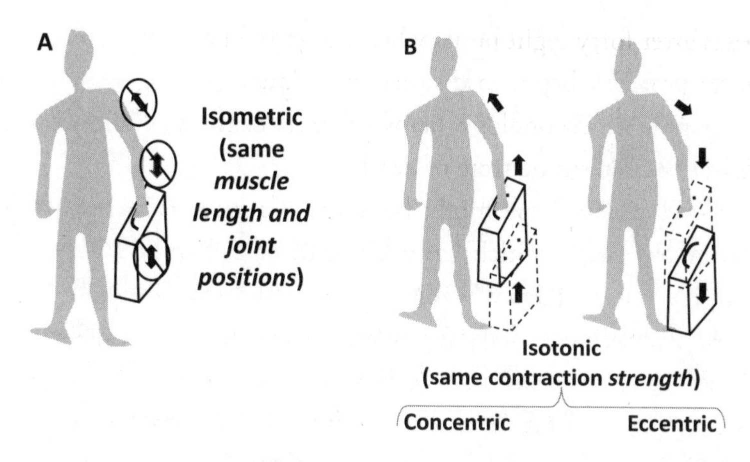

With an isometric contraction, neither the muscles, the joints, nor the load moves. Isotonic contractions are of two types: a concentric isotonic contraction lifts the load; an eccentric isotonic contraction lowers the load.

is the opposite. The muscles are still contracting but they lengthen as the suitcase descends. The same thing happens when you walk downhill. Your calf muscles are contracting just enough to keep you from falling forward as they are lengthening.

To understand what is happening with the myosin and actin during the different types of muscle contractions, imagine the Viking crew members in three situations. They might be rowing against a strong current, just holding their own. These are *isometric* contractions, in which the myosin "oars" are neither advancing nor retreating along the actin filaments. Or the crew can row and advance their ship toward a fixed landmark such as an island. These are *concentric* contractions, in which the myosin "oars" swing forward and advance on the actin filaments. Or the ship might encounter an extremely strong current that pushes it backward despite the crew's strenuous efforts. These are *eccentric* contractions, in which the myosin filaments slowly are drawn back from fully engaging with the actin filaments despite their best efforts to advance.

Recruiting the appropriate number of muscle fibers to perform a given task is a learned response, yet sometimes we get tricked or even hurt. Have you ever started to pick up a suitcase and been surprised to find it markedly heavier or lighter than you had prepared your muscles for? This may have led you to suddenly perform a surprise concentric isotonic contraction as the suitcase flew into the air or a disturbing isometric contraction when it didn't move at all.

Here is an extreme example. A competitive weight lifter will routinely drop the barbell to the mat after maximally raising it, which entails a concentric contraction—the triceps on the back of the upper arm is contracting and shortening to straighten the elbow and elevate the bar overhead. The hundreds of pounds of weight at the ends of the bar, enough to make it bow, could cause a muscle tear or tendon detachment if the athlete then tried to slowly lower the weight, which would require the triceps to lengthen while it was still resisting the weight (eccentric contraction). So lifters just make sure that their toes are out of the way and dispense with the risky eccentric contraction. At the gym, "isotonic," "isometric," "concentric," and "eccentric" are tossed around as frequently as sweaty towels, and the various forms of muscle contraction have their pros and cons with respect to weight training and conditioning (discussed further in chapter 6). Now let's look at another property of muscle that is also related to athletic performance.

Twitchiness

Skeletal muscle fibers differ according to whether they are suited better for immediate, powerful demands or for sustained, moderate contractions. Those fibers capable of producing sudden, explosive contractions are categorized as *fast twitch* because the myosin is capable of speeding along the actin at a rate of up to 70 stroke-recover steps per second. By contrast, *slow twitch* fibers can repeatedly contract for hours, but at rates of no more than 30 steps per second.

Let's talk turkey. A Thanksgiving roaster demonstrates both types of fiber. Turkeys use their breast and wing muscles to flap their way onto tree branches. This requires sudden propulsion, the kind provided by fast-twitch muscle fibers with the immediately available ATP to energize them. By contrast, while leisurely scratching and strutting, the birds contract the slow-twitch fibers that populate their leg and thigh muscles, relying on glycogen and oxygen for energy. Hemoglobin in the blood carries the needed oxygen to the surface of the muscle, where myoglobin, a similar reddish molecule, takes over the transport. When heated, myoglobin turns brown—hence the description "dark meat" for cooked turkey (and chicken) thighs and legs, the parts that are rich in slow-twitch muscle fibers and contain high concentrations of myoglobin. The breasts and wings are "white meat," full of fast-twitch muscle fibers, which have less myoglobin.

In contrast, duck breast is dark meat because it is principally populated by slow-twitch muscle fibers, useful for this migratory bird's prolonged flight.

Some people naturally have more fast-twitch or slow-twitch muscle fibers than others; and to capitalize on what nature provided, athletes gravitate to physical activities that best highlight their attributes. Proper training techniques can then accentuate which type of muscle fiber is already in abundance and preferred. People with a preponderance of fast-twitch muscle fibers make good sprinters, jumpers, and power lifters.

✦✦✦✦

TO MARVEL AT the capabilities of fast-twitch muscle fibers, I look at the ceiling in my study. It is 8 feet, 4 inches high, which was probably standard when our vintage house was built. The molding drops down about 4 inches, and on tiptoe I cannot reach it. I look up again and gasp at the fact that in 1993, Cuban Javier Sotomayor propelled his whole body higher than I can reach to set the world high-jump record. That is a record for the ages, because in the ensuing decades, nobody has explosively fast-twitched their muscles as high.

Other athletic endeavors that require a superb abundance of fast-twitch muscle fibers include sprinting, pole vaulting, and slam-dunking.

Furthermore, consider the Cossack dancers who spring repeatedly from a full-squat position on one foot while kicking their other foot out front, first one side, then the other. Whatever burning agony their quads and glutes must be feeling beneath the performers' perpetual smiles, their abundant fast-twitch muscle fibers are hard at work.

Those with a preponderance of slow-twitch fibers will find their efforts better rewarded in endurance activities such as rowing, cross-country skiing, and long-distance running. One such record is awe inspiring and will likely stand for decades, perhaps longer. Even more remarkably, American Denise Mueller-Korenek accomplished it in middle age. She had won 15 road, track, and mountain biking national championships before retiring at age 19. At age 36, she resumed cycling, and in 2018, at the Bonneville Salt Flats, she set the world's paced-bicycle speed record. For this, a race car equipped with a special cowling shields the rider from the category 5+ hurricane winds. Riding in this slipstream, Mueller-Korenek maxed out at 183.9 miles per hour, breaking the existing record, held by a man, by almost 17 miles per hour. Imagine that. She went far faster on a bike than I have ever experienced in a car or even during takeoff or landing in a jet plane.

To a limited extent, conditioning can change fast-twitch, ATP-demanding muscle fibers into slow-twitch glycogen/oxygen consumers and vice versa, but I don't expect that we will see Javier Sotomayor types winning mountain stages in the Tour de France or Denise Mueller-Korenek types performing Cossack dances.

✦✦✦✦

Throughout this chapter, I have tried to explain the reciprocating action of actin and myosin by using analogies and examples related to the type of muscle with which everyone is most familiar. It is the type that attaches to our bones—skeletal muscle. Next, I want to describe how the ever-so-minuscule actin-myosin units can scale up to produce visible and palpable skeletal muscles that can generate a smile, a home run, or both. And scale up they do. Skeletal muscle constitutes about 40 percent of our body mass.

Chapter 3

SKELETAL MUSCLE

SKELETAL MOVEMENT COMES ABOUT FROM ONE BONE MOVING in relationship to another. The intersection between the bones is, of course, a joint. Spanning the joint from bone to bone are ligaments, which are tough, usually thin, fibrous straps that keep the joint from flopping around uncontrollably. For instance, consider the ligaments on the sides of the knee. Their position and bone-to-bone attachments allow the knee to straighten and bend but prevent it from wobbling side to side or flying apart.

Cord-like tendons consist of the same tough fibrous material as ligaments. Whereas ligaments connect bone to bone across a joint,

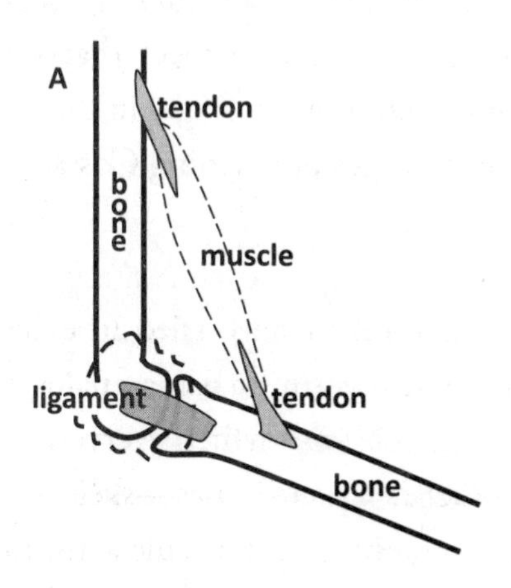

A. Tendons connect muscle to bone. Ligaments span joints and connect bone to bone. As it contracts, a muscle shortens and moves the joint that it spans.

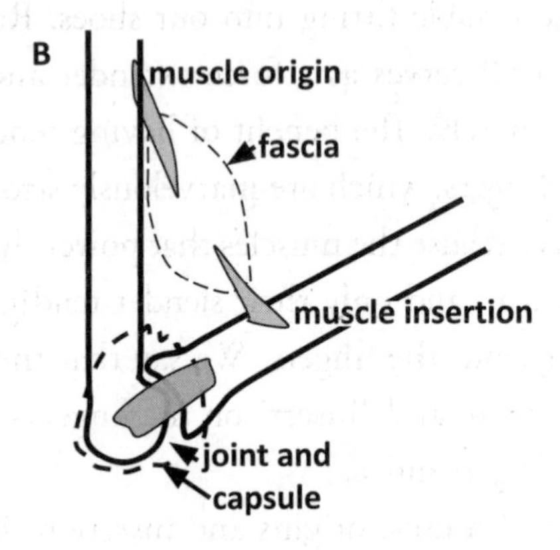

B. The muscle's attachment on the bone that does not move when the muscle contracts is the muscle's "origin." The muscle's attachment on the bone that moves is the muscle's "insertion." The joint capsule is flexible, encloses the joint, and changes shape as the joint moves. Fascia is a tough, filmy layer that surrounds a muscle.

tendons also cross joints but connect *muscle* to bone. Hence, on muscle contraction, this muscle–tendon pairing opens, closes, or rotates the joint. When you forcefully straighten your fingers, you can see the tendons on the back of your hand tenting the skin en route from the muscles in your forearm to bones in your fingers. These are "skeletal" muscles because they attach to bone either directly or through a tendinous continuation.

You might ask, what's the point of tendons? Why can't muscles attach directly to bone on both ends? The reason is that muscles are bulky and become more so when they contract. If muscles themselves crossed joints, their bulk would restrict the joints' movements, just like what would happen to your knee motion if you stuffed a loaf of bread up your pant leg behind your knee. In contrast, muscles are typically situated away from the joints they control, and their forces are transmitted across the joints through their skinny tendon extensions.

For example, contraction of the sizable calf muscles allows us to stand on our tiptoes, but if the muscles themselves crossed the ankle, we would have trouble fitting into our shoes. Rather, the Achilles tendon (heel cord) serves as a force extender and allows the muscles to work remotely. The benefit of having tendons is maximally evident in the fingers, which are marvelously strong yet slender and flexible. That is because the muscles that power the fingers are in the palm and forearm, and only their slender tendinous continuations make their way into the fingers. We say that these muscles "originate" in the forearm and "insert" on the fingers—the forearm stays still while the fingers move.

The concept of muscle origins and insertions is not always quite so simple. Consider the biceps. (*Biceps*, by the way, is both singular and plural.) It originates on bones near the shoulder and inserts on a forearm bone just beyond the elbow. When the biceps contracts when curling a dumbbell, for instance, the upper arm (origin) remains stationary, and the forearm (insertion) moves, folding the elbow closed. That is the naming convention. The bone that stays still is the muscle's origin, while the bone that the muscle moves is its insertion. To do a pull-up, however, the biceps again contracts, but now the forearm remains stationary and the shoulder moves up to meet the wrist as the elbow folds closed. So now is the forearm attachment the biceps' origin and the upper arm its insertion? In such cases it is best to call both attachments just that—"attachments."

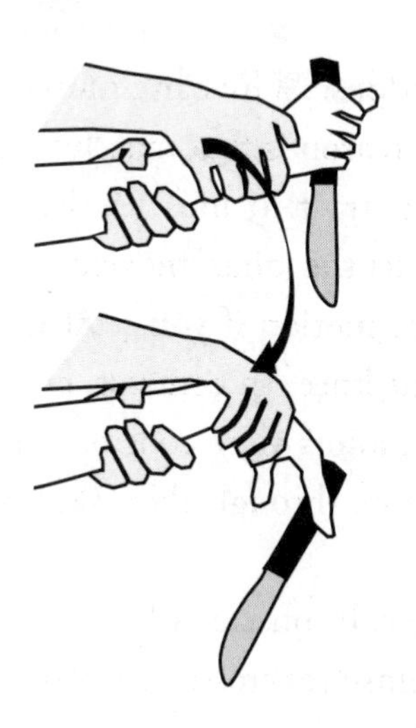

Forcefully bending the wrist down tightens the tendons that normally straighten the fingers and loosens those that normally create a forceful grip. With this maneuver, firmly grasped objects fall free without the disarmer having to grab the object or try to pry the grasper's fingers open.

Also at play is whether the muscle crosses only one joint or more than one joint between its origin and insertion. There is a limit to how much a muscle can lengthen or shorten, and if it crosses two or more joints, the position of one joint may greatly affect the muscle's performance at the other joint(s). The police take advantage of these length limitations of muscle contraction and relaxation when disarming a knife-wielding assailant without having to pry the culprit's fingers open. They merely force the culprit's wrist into a flexed position and the knife falls away. To demonstrate it on yourself (no knife necessary), first turn your palm up and make a tight fist. Now bend your fist toward your elbow crease, and then take your other hand and push the back of your hand even closer toward your elbow. Notice that you can no longer make a forceful fist. The muscles that close the fingers have expended most of their contractility across the acutely bent wrist, and they cannot shorten any more to make a tight fist. At the same time, the muscles that normally straighten the fingers are stretched to their limit across the back (convex) side of the wrist, and they cannot stretch any farther to allow the fingers to curl more tightly.

Another example is the hamstring muscles on the back of your thigh. They originate on the back of your pelvis and insert on the back of your leg just below the knee. They are therefore positioned to straighten your hip and bend your knee, but if you have tight hamstrings, they may not be able to stretch enough to allow you to do both

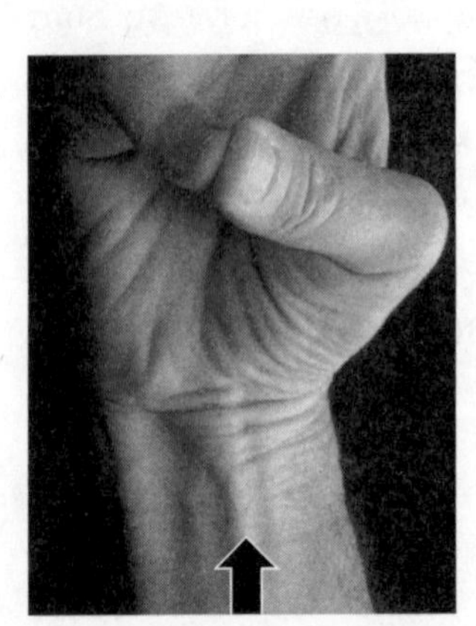

The palmaris longus tendon tents the skin on the forearm when the small finger and thumb are pinched together while bending the wrist slightly toward the elbow. Its presence, however, is variable.

simultaneously. Try it. Sit forward in a chair. Your hip is flexed. Now lift your foot off the floor by completely straightening your knee. You will probably notice a pulling sensation on the back of your thigh, and you might find yourself leaning back, which straightens your hip and gives the hamstring muscles a bit of slack near their origins to be taken up by the straightened knee near their insertions.

Voluntary? Not Always

The aforementioned hamstrings, calf muscles, and biceps, extended by tendons, are classical skeletal muscles because they attach to bone at each end. To add to the marvel and complexity of muscle anatomy, some so-called skeletal muscles attach to bone on just one end and to fibrous tissue, cartilage, or skin rather than to the skeleton on the other. For instance, there are four worm-shaped muscles in each hand and foot that originate from tendons rather than bones. Here's another example. If you forcefully pinch the tips of your thumb and small fingers together while bending your wrist toward the fold in your elbow, you are 85 percent likely to see a pencil-diameter tendon tenting the skin on your forearm. That muscle originates from bone near the elbow and inserts into fibrous tissue in the palm, not bone.

Many facial muscles have only one bony attachment. Skin serves as the other. How else could people smile and frown, and for a few, even wiggle their ears? One of the muscles that surround your eye, the orbicularis oculi, has three parts, two of which have no bony attachment at all. The same goes for the tongue. Four of its muscles attach to bone only at one end; the other four have no skeletal connection at all. It is an identical situation for the muscles controlling our vocal cords. They both originate and insert on cartilage. Yet all these muscles are deemed to be "skeletal" muscles because they are voluntary, meaning we can contract them and move our elbow, knee, or lip when we want to.

Many "voluntary" skeletal muscles are out of our control at least part of the time, and mercifully so. Consider again the orbicularis oculi surrounding your eye. According to the need for speed, you can voluntarily contract it and wink, but it also contracts reflexively when a bug flies into your eye, faster than if you consciously willed the muscle to contract.

Coughing and breathing are similarly under dual control. In the past few minutes, you probably have not been thinking about your rhythmically contracting diaphragm. But now you are. From chapter 1, remember Andreas Vesalius, one of the first human anatomists? He opined, "If respiration did not depend on our own will and impulse, we would not be able to make speeches of long duration, and this would adversely affect our quality of life." (Was he being funny?)

At least one skeletal muscle is entirely involuntary in its action. It reflexively protects the inner ear from receiving offensively loud sounds by damping large oscillations passing through the middle ear bones. Incidentally this muscle, the stapedius, is, at less than a quarter of an inch long, the smallest one in the body.

Other so-called voluntary muscles behave situationally. Shivering is an entirely involuntary response to cold or fear, and our teeth chatter whether we want them to or not.

Sometimes too we unwittingly and perhaps against our best interests contract voluntary muscles that convey thoughts—body language. This includes not only facial expressions and eye movements but also posture, hand gestures, and use of space. Mentalists are adept at reading body movements, as are successful poker players. They read the thoughts of others while remaining perfectly poker-faced themselves. Despite one's best efforts at concealing thoughts, however, sometimes a flicker across a person's face lasting no more than a fifth of a second reveals emotion. Such brief muscle contractions, which the owner is not likely aware of and which casual observers would miss, are called micro-emotions, also known as tells. They are a study unto themselves, and controversy exists whether customs-control agents, for instance, can learn to spot these fleeting glimpses into a smuggler's mind. Although I am not sure that the vocal cords express micro-emotions, certainly one's voice will crack when one is stressed, so maybe TSA agents should ask travelers to sing a little.

Oddities

I have come to enjoy skeletal muscles' vagaries regarding their attachments and control. It makes them interesting rather than confusing. I hope you too can appreciate their variations in form, function, and control. Consider it this way. People look different—tall, short, light, dark, skinny, heavyset, curly hair, no hair, asymmetric faces. That is why it is interesting to look at people, and sometimes the findings are hard to accurately name and classify. How boring it would be if all of us were cookie-cutter images of one another. Well, anatomical variations and ambiguities do not stop at the skin. There are myriad internal variations as well. They occur in all organ systems and keep anatomists and doctors on their toes.

For instance, about one in every 10,000 people have their internal organs entirely flipped from side to side. The heart is on the right, and the appendix, usually on the right, is on the left. By the same token, a skeletal muscle can be present in one limb and not the other. So, if somebody sees you reading this book and asks, "How many skeletal muscles are there?" sigh and answer "about 650," with the emphasis on *about*. Here are some common skeletal muscle variations that account for the ambiguity.

On page 36, I showed you my palmaris longus tendon and told you how to check to see if you have one in your forearm. The incidence of its presence varies by ethnicity. Roughly 85 percent of Caucasians have it in both forearms, 8 percent in just one forearm, and 7 percent in neither. Only 5 percent of African Americans and 3 percent of Asians are missing it. The good news is that the people who do not have a palmaris longus are fully functional. Hand surgeons know that the palmaris longus does not endow its owner with any special powers and consider it expendable. We at times "borrow" it to substitute for a missing critical tendon elsewhere. (I elaborate on this in chapter 8.)

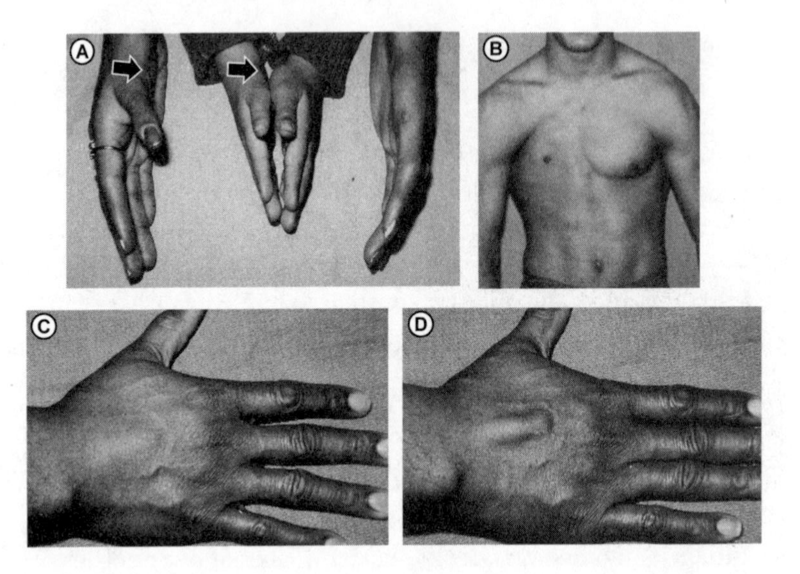

A. From birth, both mother and daughter are missing the fleshy muscles at the base of their right thumb. The mother's left thumb failed to develop altogether, whereas her daughter's left thumb is normal.
B. This high school wrestler was unimpaired by the congenital absence of his right breast muscle (pectoralis major).
C. Although short muscles are normal on the top of the foot, they are rare on the back of the hand and may be mistaken for tumors.
D. The ability to voluntarily contract this muscle confirms its nature.

The same holds true for the plantaris muscle, which runs down the back of the leg with the calf muscles. The plantaris is absent in about the same proportion of individuals as the palmaris longus. It is equally useful as a donor tendon when it is present. With less frequency, other muscle anomalies occur; see the photographs on the next page of several that I have encountered in my practice.

Reflexes

I have mentioned several reflexes. None is better known than the knee jerk. You can willfully contract your thigh muscles and straighten your leg; or you can relax, cross one knee over the other,

To demonstrate the Jendrassik maneuver, the subject isometrically contracts upper extremity muscles by trying to pull his hands apart while the examiner tests his knee-jerk reflex.

and have somebody tap on the tendon just below your kneecap. (A wooden cooking spoon works as an impromptu reflex hammer.) Whoopee, the knee momentarily straightens a bit. That is a spinal cord reflex. The light tap stretches the muscle ever so slightly and activates stretch receptors residing inside. They send electrical messages up the limb to the spinal cord, which immediately fires back a message for the thigh muscle to contract. The brain is shorted out.

Well, the brain is *usually* not involved. In the late nineteenth century, Hungarian physician Ernő Jendrassik described another vagary, which is now known as the Jendrassik maneuver. Before you have your assistant tap your patellar tendon, clench your teeth and try to pull your grasped hands apart. Doing so puts the nervous system on red alert and accentuates the jerk. Be careful not to kick your assistant involuntarily.

The stretch receptors are one of two specialized monitors embedded in skeletal muscles or in their tendons. As exemplified by the knee-jerk reflex, the stretch receptors, buried within the muscle, protect a relaxed muscle from suddenly being yanked into an elongated position and possibly damaged. If the stretch occurs slowly, however, say over seven to ten seconds, the stretch receptors are not activated. They sense that nothing terrible is in store and do not reflexively halt the elongation. Hence, slow steady stretches, in yoga, for instance, yield progressively more flexibility. Think of the stretch receptors as sleeping guard dogs. If you shout at them, they bark. Softly sweet-talk them, and they will want their bellies rubbed.

In addition to protecting relaxed muscles from suddenly being overly lengthened, the stretch receptors contribute greatly to "position sense," which neurologists consider to be our sixth sense. The well-known five senses inform us about our surroundings, whereas position sense tells us, independent of eyesight, our body's orientation with respect to gravity and where our body parts are situated relative to one another. Thereby we can walk in the dark without falling over and can know where our fingers are on a guitar or piano keyboard without looking at them. This finely tuned system, however, often cannot keep up with rapidly growing teenage bones and muscles, and its lag in adapting to the elongated limbs accounts for awkward movements. At any age, alcohol impairs these positional monitors. The police know this and field test for sobriety by asking suspects to close their eyes and try to touch their nose with their index finger or having them walk heel-to-toe in a straight line.

Working in conjunction with the intramuscular stretch receptors is another set of monitors, the Golgi tendon organs. They are embedded in tendons and sense tension. When they recognize that a heavy weight is greater than what the contracting muscle can manage, via a reflex the Golgi tendon organs force the muscle to relax, protecting it against tears but also perhaps causing the owner to drop an impossibly heavy object. The Golgi tendon organs also discern heft and, for instance, can reveal how much paint remains in a spray can.

Hysterical Strength

Physiologists calculate that ordinarily we use only about 60 percent of a muscle's theoretical capacity before the intramuscular stretch receptors and the Golgi tendon organs begin sounding alarms. This causes the brain to stop sending "contract" messages. The muscle reflexively relaxes and avoids tearing itself or ligaments or breaking a bone. (Yes, bones can break due to extreme muscle contraction.) This "too much" message apparently gets sent sooner in some people than in others, and athletes have the capacity, either innately, through training, or both, to ignore the first call to surrender and can contract a muscle to about 80 percent of its theoretical capacity.

Can muscles do even more? Newspaper headlines say they can. "Man lifts car off pinned cyclist." "Boy's strength muscles car off grandpa, saves life." These heroes are ordinary folk, untrained and entirely unprepared, who in an instant of great need perform seemingly impossible feats of strength, called hysterical strength. How is this possible? Sure, video clips exist of hulking and grunting behemoths rolling small cars over, but these hulks are thoroughly weight trained, and only their successful effort, perhaps one out of many, made it onto YouTube. That is probably not the case, however, behind the headline "Oregon man pinned under 3,000-pound tractor saved by teen daughters." In such instances the rescuer(s) may have given the victim enough room to wiggle free by tilting the overturned vehicle as if it were a table with a short leg.

In other occurrences, it is likely that the hero may have lifted one wheel to give the victim room to escape. In the gym, this maneuver is called a dead lift, performed by starting with bent hips and knees and straight back, grabbing the barbell, and standing straight up. The world record for men is over 1,100 pounds and for women almost 700 pounds. In a life-or-death situation, how much could a normal person dead-lift? Emotional stress, such as seeing a child trapped beneath a car, having a trainer or drill sergeant screaming in your face, or hearing thousands of maniacal home-team fans hollering their heads off, can reset the alarm threshold. Adrenaline can certainly enhance the force of muscle contraction, but the call for superhuman strength is likely to come and go before this hormone, released from the adrenal glands, would have a chance to circulate and influence muscles.

The phenomenon of hysterical strength is difficult to study in a controlled manner since the ethics of simulating a do-or-die scenario would be difficult to justify. One pair of researchers did, however, come close. Their results support the idea that psychologically induced inhibitions limit expressions of strength. Seated subjects held a grip meter and were instructed to forcefully squeeze it at regular intervals. Randomly one of the researchers, standing behind the subject, would fire a starter's pistol into the air. On average, the next squeeze was 7 percent greater than baseline. When the researchers asked the subjects to shout as loudly as possible while squeezing, the grip strengths averaged 12 percent greater than baseline.

This is perhaps why warriors, at least in movies like *Braveheart*, holler blood-curdling rage as they rise to confront the enemy. In addition to potentially intimidating their quarry, the emotional fury most likely raises their own muscles' functional capacity from 60 percent to closer to 80 percent without setting off their Golgi tendon organs and stretch receptors to inhibit forceful contractions. The same also seems to work for modern-day tennis players, martial arts practitioners, and weight lifters, many of whom forcefully exhale with a grunt just at the instant of maximum physical effort. (Some gyms prohibit such primal sounds.) Researchers have shown that college tennis players increase the velocity of both their serve and their forehand volley by 5 percent when they grunt at the instant of impact. The grunt not only improves their game, it may also distract their opponent, which many would consider foul play.

Another neuromuscular oddity and possible party trick, depending on one's idea of fun, is a demonstration of the Kohnstamm phenomenon. German neurologist and psychiatrist Oskar Kohnstamm described it in 1915, although I wonder if children haven't known about this effect for eons. Stand in a doorway and press the backs of your hands against the frame for about thirty seconds, as if you are trying to widen the doorway. Then step away and relax. Magically your arms will float away from your sides.

The phenomenon works for other sets of muscles as well, and indi-

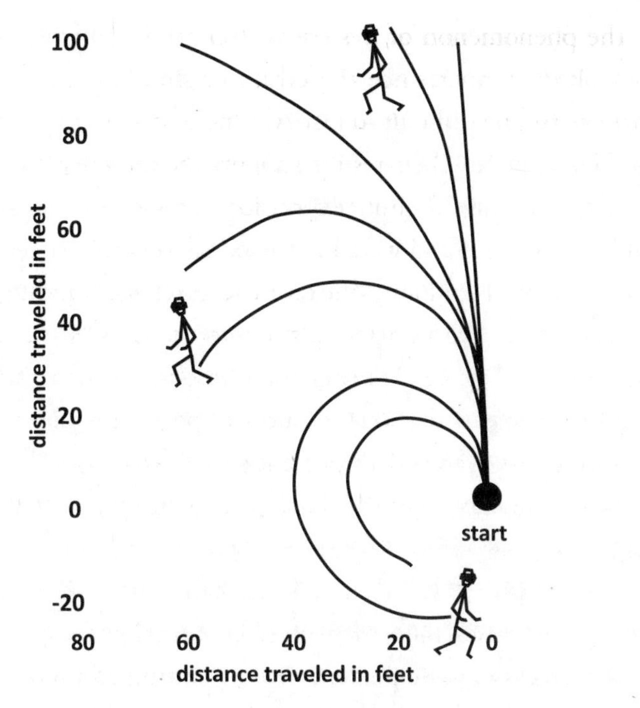

These are representative walking paths taken by blindfolded subjects across an open field after having resisted a trunk-twisting maneuver. They demonstrate the presence of the Kohnstamm phenomenon to different degrees.

viduals are variably affected. For example, one group of researchers blindfolded standing subjects at the edge of an open field and asked them to twist their torsos to the left. Researchers seated behind them held their hips and resisted the twist. After 30 seconds, the investigator instructed the blindfolded subjects to walk forward in a straight line. Most subjects involuntarily veered to the left from between 45 and 270 degrees, demonstrating the Kohnstamm phenomenon. Nobody is absolutely certain what causes this interplay between nerves and so-called voluntary muscles.

Then there are the cremaster muscles. They are responsible for raising and lowering the testicles to keep them in a temperate zone so the sperm they house are neither chilly nor overheated. Men have minimal voluntary control over this thermostatic elevator, and to my knowledge the effects of climate change on the cremasters' function are yet to be tested.

Rather than the mind controlling the body, in certain instances the body can control the mind. Researchers have shown that when you hold a pencil between your teeth, it forces you to smile, and doing so is likely to make your thoughts happy ones. Conversely, if you pinch the pencil between your upper lip and your nose, your facial muscles force a frown, and your thoughts are likely to turn negative. This is called embodied cognition.

In one study, subjects were timed while they attended to a task with their arms either crossed or uncrossed. Those with crossed arms endured longer, suggesting more resolve. On a similar note, researchers have found that assuming a confident Superman stance for two minutes before engaging in a stressful activity improves one's mood. You might draw some stares when spotted outside a courtroom or boardroom in superhero stance with your arms crossed and holding a pencil between your teeth, but if it works, go for it. Embodied cognition, a fertile area of research and debate, further demonstrates the complex interplay of mind and muscle and our incomplete understanding of it.

Central Control

We know now that electrical impulses racing along nerve fibers transmit messages back and forth between brain and muscle. This awareness began with a chance discovery made by Luigi Galvani and his wife, Lucia, in 1780. It flew in the face of over 2,000 years of speculation and dogma. The speculation started in about 400 BCE with Hippocrates, who thought tendons and nerves were one and the same. Aristotle some 50 years later set that opinion aside, and 50 years after that, Erasistratus made another mark on science. Remember, he is credited with performing the first systematic human dissection. In addition, he traced nerves to the brain and posited that "psychic pneuma" coursed through hollow nerves, ballooning the muscles and causing them to contract. Galen held influence beyond all measure and espoused Erasistratus's theory. Galen promoted this "balloonist theory" of nervous fluid powering muscle, and his writings on this topic along with others held sway for centuries. Overall, Galen wins my award for supporting the longest-held erroneous speculations in the history of biological science.

An inkling that Galen was wrong regarding the balloonist theory came in 1666, when a Dutch biologist, Jan Swammerdam, demonstrated in an elegant experiment that contracting muscles did not balloon. Then the Galvanis (Luigi, of course, the times being what they were, received all the credit) showed that when they shocked the hind limb nerve of a dead frog, the leg would jump, demonstrating animal electricity, later known as galvanism. This eradicated several thousand years of speculation regarding "psychic pneuma" and "animal spirits" and led to the formation of a new science—electrophysiology.

Today, tests for electrical activity in the brain (electroencephalography), in nerves (electroneurography), and in muscles (electromyography [EMG]) are routine. Compared to electrocardiograms, where surface electrodes at the wrists and ankles and directly over the heart suffice, precision EMGs routinely require placement of a needle electrode directly into the muscle of interest. At rest, healthy skeletal muscle is electrically silent but then generates the visual equivalent of loud static on the monitor during maximal muscle contraction.

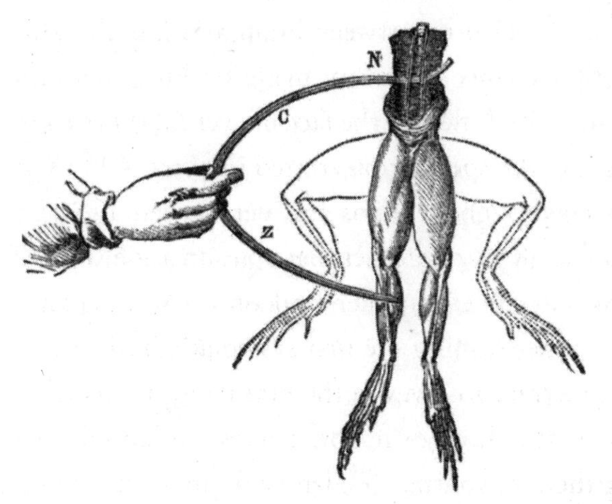

This 1859 drawing recognizes Luigi and Lucia Galvani's discovery 80 years before. They noted that stimulating the leg of a dead frog with electricity made its muscles contract.

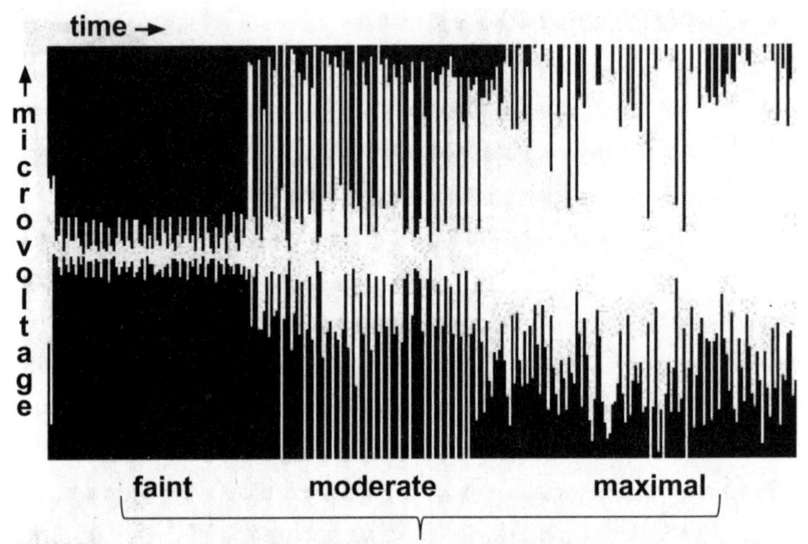

This EMG tracing shows several seconds of electrical activity in a skeletal muscle during faint, moderate, and maximal voluntary contraction. With progressively more forceful contraction, the already activated motor units fire more rapidly, and more muscle fibers are activated. Together this causes increases in both the frequency of electrical discharges (waveforms jammed together horizontally) and in their intensity (waveforms of increasing voltage).

Evidence of electrical activity at rest and altered wave forms seen during activity can indicate problems at different sites. This could be within or near the spine (for example, spinal stenosis, amyotrophic lateral sclerosis), in the nerve on its way to the muscle (for example, diabetic neuropathy, carpal tunnel syndrome), where the nerve contacts the muscle (for example, botulinum toxicity), or within the muscle itself (muscular dystrophy). Do you remember the mention, in chapter 2, that by grimacing forcefully you can hear a low-level hum created by myosin molecules stepping along actin filaments? Researchers are testing "phonomyography" to see if they can obtain better, or different, information about a muscle by listening to it compared to using electromyography.

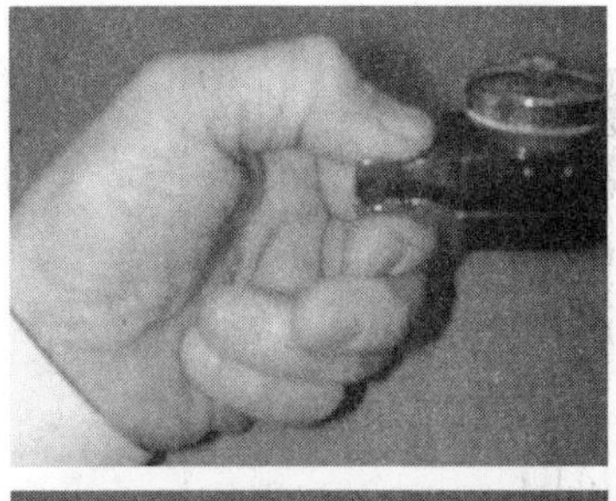

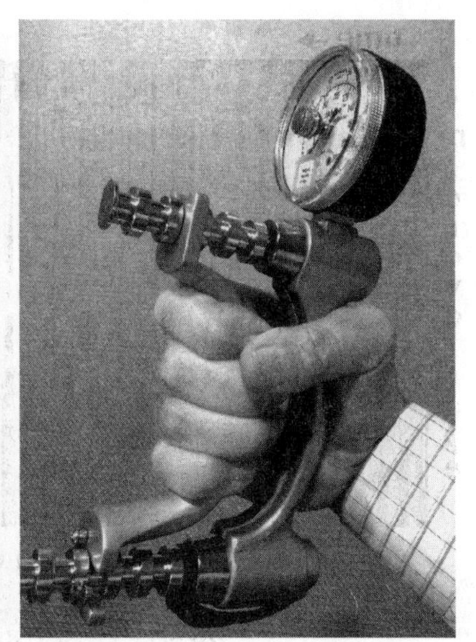

Upper left: A pinch meter is positioned to measure "side" pinch between the thumb and the index finger.
Lower left: Here, thumb-to-index "tip" pinch is measured.
Right: The grip meter can be adjusted to measure maximal strength in hands of all sizes.

Prior to ordering an EMG study, which some patients indicate is not an entirely comfortable experience, and perhaps in lieu of it, doctors commonly perform tests for muscle strength. The assessments range from simple qualitative tests ("Squeeze my fingers," "Can you stand up from a sitting position by yourself?") to complex ones using electromechanical devices known as dynamometers, which put numbers on the results. Some dynamometers are handheld (grip meters, pinch meters). Others are floor-based, heavy, and adaptable for testing various muscle groups in the trunk and limbs. A few are uniquely designed for specific tasks.

For example, the efficiency of swallowing in patients who have sustained strokes can be monitored before and after tongue-strengthening exercises by using the Iowa Oral Performance Instrument. This dynamometer fits in the mouth and measures the force of the patient's tongue pressing against the roof of the mouth. And women, particularly those who have experienced childbirth with subsequent weakness of their pelvic floor musculature, can get a precise picture of the progress of their Kegel exercise program with the use of a perineometer.

Monitoring Fitness

Along with strength, other measurements that help define fitness are muscle mass, lean body mass, and total body fat content, and all four determinants are variants of one another. Basically, muscle is good, blubber is not. Periodically ascertaining the percentage of each and monitoring changes in their proportions can measure the road to fitness.

Three tests can be easily done at home. The tape-measure test requires recording the neck and waist circumferences for men and neck, waist, and hip circumferences for women. You can enter these measurements along with your body weight and height in an online calculator (search for "body fat calculator"). Repeated periodically, any change in the body's proportion of fat versus muscle becomes evident. This tape-measure test is the least complicated but also the least accurate, in part because of inconsistency on how tight the measuring tape is pulled.

The next home test is pinching a fold of skin and measuring the thickness of the intervening fat layer with calipers, which are economical and readily available to purchase online. The instructions that accompanied the calipers I bought had me take just one measurement, of an abdominal fold. I then entered that number onto a sex- and age-specific chart, which yielded my percentage of body fat. Another method takes skinfold measurements from three, seven, or nine body areas and inserts those numbers into formulas.

The third home test requires a special bathroom scale or handheld device, both of which are also available online. The scale sends a faint electrical current up through one foot and measures what comes back down into the other foot. The handheld device does the same, going across the body between the hands. Fat resists electricity more than muscle does, so higher conductivity indicates more muscle. It may be the least accurate of the tests since it is highly dependent on hydration level and how much time has passed since one has eaten a meal or exercised.

Of available lab-based tests, several compare the body's volume to its weight. Muscle weighs more than fat, so for two people of equal volume, the one who weighs more is more muscular. Volume can be determined by how much water or air a totally immersed body displaces. Fill a special tub to the brim, totally submerge the minimally clad subject, collect and weigh the overflow, and calculate its volume. Or enclose the subject in an egg-shaped capsule, the Bod Pod®, which measures the volume of enclosed air with and without the subject inside. Or perform an electronic scan of the subject and create a 3D full-scale reproduction, from which the computer calculates the volume.

Another way to determine body fat is to obtain a dual energy X-ray absorption (DEXA) study, which differentiates fat from muscle during a several-minute head-to-toe scan. The amount of radiation absorbed is minuscule, less than what we absorb each day just walking around. (This scan is different from the more sophisticated test by the same name that determines bone density.) This type of DEXA scan is regularly used by nutritionists, gerontologists, cardiologists, and sports medicine specialists to periodically monitor body composition.

Each test has its pros and cons. One requires getting wet. The tape-measure skinfold tests require the least equipment. Several will produce different results according to the subject's level of hydration. If you are interested in monitoring your level of body fat, which is one indicator of fitness, pick one method and stick with it. Test at the same time of day and with the same level of hydration.

Another indication of muscle fitness comes from measuring one's

VO_2max (*vee-oh-two-max*), which is the maximum volume of oxygen one's muscles can use when going all out with an endurance activity. It considers efficiency of respiration, cardiac output, oxygen-carrying capacity of the blood vessels, and the muscles' ability to extract and use oxygen. The accurate determination of VO_2max takes place in a laboratory with the subject on a treadmill or exercise bike and with monitors for heart rate and oxygen consumption in place. That is a lot of equipment, and not everybody should challenge their heart to beat maximally, so there are formulas (available on the Internet) that estimate VO_2max by considering age, resting heart rate, and ending heart rate after a one-mile jog or fast walk.

Average VO_2max scores (measured in milliliters of oxygen consumed in one minute per kilogram of body weight) for normal human beings in their twenties are 48 for men and 38 for women. They then decline by about 10 percent per decade, which parallels the predictable decade-by-decade loss of skeletal muscle mass. This explains why sooner or later you may notice that ascending several flights of steps leaves you breathless. Hard training can reduce this inevitable decline by about half. The numbers probably don't mean much unless you compare them to the VO_2max scores of elite athletes.

Successful endurance athletes likely gravitate to sports like distance running and cross-country ski racing because they naturally have a great capacity to get volumes of oxygen to their muscles, which are highly populated with slow-twitch fibers. Then they train to enhance and capitalize on those characteristics. Here are the VO_2max scores for a sampling of elite endurance athletes. It should not be surprising that their scores are roughly double that of "ordinary" human beings.

VO_2max

98 Oscar Svendsen, age 18, cyclist, Norway

93 Greg LeMond, cyclist, US, winner Tour de France, ages 25, 28, 29

84 Steve Prefontaine, runner, US, set every record 2k–10k, age 24
84 Lance Armstrong, cyclist, US, seven-time winner Tour de France, ages 28–34
81 Jim Ryun, runner, US, first high school miler to break 4:00 minutes; set world record for mile 3:51.1, age 20
79 Joan Benoit, 1984 Olympic marathon gold medalist, age 27
76 Michael Phelps, swimmer, US, record, 23 Olympic gold medals

This chapter has focused on the basics of skeletal muscles. To support and greatly enrich the body's capacity for movement, two other types of muscle have critical roles. Both types are entirely nonskeletal and are almost always involuntary.

Chapter 4

SMOOTH MUSCLE

DID YOU EVER SEE THE MOVIE *PSYCHO*? WHEN YOU SAW THE creepy scene at the end, did it give you goose bumps? If you were at all aware of tiny bumps all over your skin, you may have wondered why we get them.

Muscles cause them. These wisps of muscle connect the bases of hair follicles to the undersurface of the skin. When they contract under conditions of cold or fear, they lift and straighten the follicle, which creates a bump on the skin and stands the hair fiber on end. In furry animals this thickens their coat and provides additional cold protection. Raised hackles—erect neck and back hair—make the owner look larger and more intimidating, so it is likely best to back away from a canine, feline, or cervid (such as a moose) displaying this automatic reaction. For humans demonstrating the phenomenon, it may suffice to offer them a sweater, a drink, or both. For your trivia library: eyebrows have this feature, eyelashes do not.

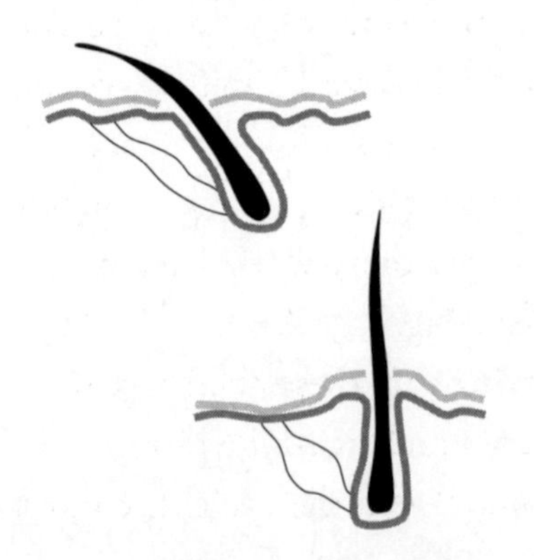

A wisp of smooth muscle connects the hair follicle's base to the skin. On contraction, the muscle lifts and straightens the follicle, raising a bump on the skin and standing the hair straight up.

This behind-the-scenes tissue containing contractile filaments of actin and myosin is called smooth muscle, but it got its name not because it is suave and confident but rather from how it looks under the microscope. (More on this later.) Goose bumps are visible and dramatic, but most of the body's smooth muscles are located deeper within and are involved in transporting gases, liquids, and solids. Our bodies are full of tubes for such purposes; to regulate the rate of transportation, smooth-muscle fibers in the tubes' walls constrict and relax to make the passageways intermittently smaller and larger. The tubes come in all sizes. Portions of the large intestine are 2.5 inches in diameter, the aorta is about half that, and the trachea slightly less. On the narrow end of the spectrum, 24 lymphatic vessels would fit inside a standard drinking straw. (Capillaries are even smaller. Nearly 200 of them would fit, but they do not contract.)

Regarding length, an artery, although it may change names along its path, is about 4 feet long by the time it courses from the heart to the toes. The award for the body's longest tube, however, goes to the small intestine, which is well muscled over its entire 22-foot course.

Furthermore, these contractile tubes are at work day and night and contribute critically to all of life's vital functions—movement, reproduction, sensitivity, growth, respiration, excretion, and nutrition—in short, the memorable MRS GREN. The organ systems responsible for these respective functions are muscular, reproductive, skin, endocrine, cardiorespiratory, urinary, and digestive. Most of the time, we are only vaguely aware of these systems and their functions, and even when we are aware, they are generally beyond our voluntary control. This might be a blow to a control freak's ego, but it is better that these muscles work automatically. Otherwise, we would fill our entire day making vital commands such as "Churn lunch," "Don't let me get dizzy when I stand up," "Cool me off," and "Keep the urine moving."

The muscles responsible for carrying out these vital functions aren't entirely restricted to tube shapes and hair-follicle wisps. In the introduction, I described those muscles in the eye that control the size of the pupil and the shape of the lens. If you are reading rather than listening to this book, those muscles are working right now—and wouldn't it be annoying if you had to constantly and consciously adjust them?

Autonomy

The nerves that control smooth muscles constitute the autonomic nervous system and are almost entirely separate from the nerves in the voluntary system, which includes the nerves that are presently holding your eyes open. The autonomic system consists of two parts, sympathetic and parasympathetic. I find neither name particularly meaningful, so it is easier to just understand each portion's effect on smooth muscles.

Think of the sympathetic portion as the gas pedal in a car and the parasympathetic portion as the brake. If a quick response is needed—say, to escape from a burning building—acceleration is required. The sympathetic nerves step on the gas to dilate the pupils (to see better at a distance), speed up the heart (to circulate more blood), and dilate lung airways (to capture more oxygen). They also dilate arteries to muscles (to provide more energy) while constricting arteries to the gut and kidneys, shunting blood temporarily away from functions not required in emergencies. In other words, the sympathetic nerves activate fight-or-flight functions or, maybe in a panic, a freeze response.

Conversely, for resting, parasympathetic stimulations apply the brakes. They slow the heart, promote digestion and urine formation, constrict the pupils to enhance near vision, and shunt blood away from big muscles and toward the internal organs. Hence, the parasympathetics are known as the rest-and-relax or feed-and-breed nerves.

The two portions of the autonomic nervous system are more collaborators than competitors, and their balance on smooth-muscle contraction and relaxation is nuanced. For instance, together they allow us to stand from sitting without getting dizzy. And while the parasympathetics are responsible for sexual arousal, that by itself wouldn't perpetuate the species: the sympathetics join in to stimulate the smooth muscles involved in orgasm.

What's more, the autonomic nervous system is not entirely autonomous. (Like life in general, it's complicated.) Here are three examples. First, conscious techniques of relieving stress, such as deep breathing, mindfulness, laughter, and monotasking (the opposite of multitasking) can activate parasympathetic stimulation to aid resting and digesting. Second, scientists have measured skin temperatures in meditating yoga masters and noted a 7 degree Fahrenheit warming of their fingers and toes while skin surfaces elsewhere remained unchanged. They are willfully relaxing the circumferential smooth muscles in the walls of their digital arteries, which allows these to dilate and deliver more blood to the skin, which warms it. Third, swallowing a sword longer than 15 inches requires passing it through the involuntary lower esophageal sphincter (a ring of muscle) into the stomach without retching. According to swordswallow.com, the longest sword swallowed in an apparently normal person was 25 inches—well past the "involuntary" gatekeeper of the gullet. The only paper I could find in the medical literature related to the side effects of sword swallowing states, "Occasionally a sword is difficult to advance or retract, presumably because of spasm or mucosal dryness related to nervousness or soreness." That sounds to me like a panicked freeze response of the sympathetic nerves.

In addition to the autonomic nervous system's division into sympathetic and parasympathetic portions, smooth muscles are also divided according to how each cell receives the electrical messages that command it to contract. The more common way is for one nerve fiber to stimulate several adjacent muscle cells. Then these cells transmit the impulse to their neighbors, and they all contract in unison. This is an economy that produces gross movements, ones that do not require detailed oversight. It works fine in the uterus, stomach, small intestine, and bladder, for instance, where generalized, regional contractions fully suffice.

The other way smooth-muscle cells receive their electrical messages offers more nuanced control—each smooth-muscle cell has its own nerve fiber. Hence you can get goose bumps on your forearms but not your neck. The iris in your eye is under similar fine control and can respond precisely according to changes in light.

Although both sympathetic and parasympathetic fibers connect with the small intestine, muscular movement here can occur without any autonomic nerve control at all. Hunters witness this capacity while field dressing a deer. Its small bowel continues to rhythmically contract even after being completely removed from the carcass. This self-regulation also accounts for the success of small-bowel transplants from one person to another without reconnecting any nerves (details discussed later). To account for this ultimate example of smooth-muscle autonomy, the digestive system has its own nervous system. It consists of hundreds of millions of nerve fibers closely networked and controlled by thousands of tiny ganglia—nests of brain-like nerve cells, entirely beyond any willful control. Roll on, digestion, roll on.

Contractable Tubes

Of the body's contractable tubes, lymphatic ducts, veins, and arteries have one layer of smooth muscle in their walls. Here, the muscle fibers are circumferentially oriented, and on contraction they narrow the tube's diameter. The arteries resist high pressures from blood being pumped from the heart and therefore have thicker smooth-muscle layers than do veins and lymphatic ducts, which have only to resist low pressures as they return fluids to the heart.

Next in complexity are the small intestine and the upper portion of the ureter, which connects the kidney to the bladder. These tubes have a circumferential layer of smooth muscle, again to contract the tubes' diameters, and a longitudinally oriented one, to reinforce

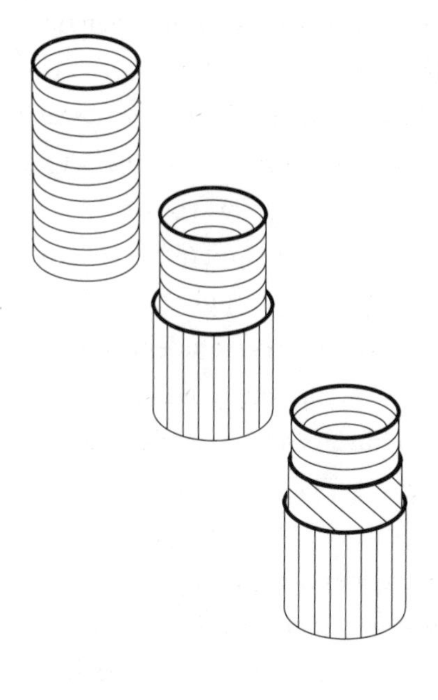

Schematics representing the orientation of smooth muscle cells in tubular structures. Left: Arteries have one layer of smooth muscle, which is oriented circumferentially. Middle: The small bowel and the upper portion of the ureter have two laminations, each containing either circumferentially or longitudinally oriented smooth-muscle cells. Right: The lower portion of the ureter has an additional layer containing obliquely oriented smooth-muscle cells.

the wall and to aid the circumferential layer in moving the tubes' contents along. The stomach has a third, obliquely oriented lamination, which helps mix food and gastric acid together. The bladder and uterus have multiple smooth-muscle layers whose lamination arrangements defy easy explanation. When the variously oriented smooth-muscle layers rhythmically contract and relax, the contents of the tube, be it partially digested food, urine, or a fertilized egg, get squeezed along by a phenomenon known as peristalsis. This is akin to running your fingers up from the bottom of a nearly empty toothpaste tube. The action's efficiency is dictated by how tightly you pinch your fingers together.

Peristalsis in the fallopian tubes only narrows the passageway slightly and ensures a compression-free transit of a fertilized egg into the uterus. Peristaltic closure is about 60 percent in the small intestine, and there may be brief periods of reverse peristalsis over short segments to ensure a good mix of food with the digestive enzymes. The circumferentially oriented smooth muscles in the ureter's wall can narrow its passageway by 95 percent. This not only efficiently advances urine from the kidney to the bladder, it is also usually sufficient to block bacteria from ascending from the bladder to infect the kidney.

Controlling the Thermostat

Now that I have described the general attributes of smooth muscles and their mostly autonomous control by sympathetic and parasympathetic nerve fibers, it's time to provide some interesting details about their specific roles and look at how they can, at times, contribute to disease.

Arching out of the heart, the aorta has smooth muscle in its wall, but it does not constrict and relax enough to change the aorta's diameter. Rather, the aorta's smooth muscle makes the aorta resilient and helps it withstand the second-by-second change in pressure as the heart forcefully disgorges another several ounces of oxygenated blood into the circulation. Farther along, the smooth muscle in all the arteries and their branches, known as arterioles, receives fibers from the autonomic nervous system.

Then something special happens just before the arterioles further divide into capillaries, where oxygen and glucose jump off and carbon dioxide and other metabolic by-products jump on. The brain and intestines are special in this regard because their arterioles have ringlike sphincters made of circumferentially oriented smooth muscle that controls the flow of blood into these organs. These sphincters are controlled by the autonomic nervous system and provide additional circulatory regulation to the bloodthirsty brain and gut. Under circumstances of fight-or-flight, the sphincters in arterioles leading to the gut tighten and those to the brain relax, thereby ensuring that the brain receives the necessary nutrients to resolve the crisis. When it is time to rest and digest, the sphincters controlling circulation to the gut relax to shunt more blood there, and the brain, with less circulation, may take a nap.

The other mechanical smooth-muscle blood-flow regulators are present only in the skin, particularly in the hands and feet. Indirectly, these sphincters perform thermoregulation for the internal organs. When blood exits the heart and takes off on its roundabout journey, it passes sequentially through artery, arteriole, capillary, venule, vein, heart, lung, and back to the heart to start another circuit.

If the internal organs are chilly, the motherly autonomic nervous system constricts small arteries just below the skin, pinching off much of its blood supply to benefit the brain and organs of the chest and abdomen. Consequently, the skin of the fingers and toes may turn pale from the scarcity of blood and may even sustain frostbite in extreme situations. Conversely, if the innards are overheating, the shunts reverse roles and direct more blood through the skin's capillaries. The skin flushes, and heat radiates away before the cooler blood heads back inside to placate those temperature-sensitive internal organs.

What can go wrong with smooth muscles? Plenty, particularly the ones incorporated in tubes. Household plumbing, by comparison, is simple. Sure, your pipes at home can clog or leak, but they aren't constantly changing their diameter according to the demand, and they don't take kindly to bending, especially repetitively and over decades. Remarkable and durable as they are, the body's smooth-muscle tubes can develop problems. Let's explore some of the most common and interesting ones.

Arterial Afflictions

In some people, especially with advancing age, the natural constricting and dilating properties of arteries can slow up and the arteries become hard—*sclerotic*, in medical parlance. The most common form of hardened arteries is atherosclerosis. A German pathologist, Felix Jacob Marchand, coined the word in 1904. *Athero* in Greek means gruel, and I guess the deposits on the arteries' inner walls reminded Dr. Marchand of thin porridge. Yuck.

The development of atherosclerosis is a complex process that specialists do not fully understand, but unless you have been living on a desert island for decades, you know that the risk factors include age, high cholesterol, high blood pressure, diabetes, obesity, smoking, genetics, and an unhealthy diet. No wonder that atherosclerotic cardiovascular disease (ASCVD) is the developed world's number one killer. Arterial plaque, entirely different from what dental hygienists scrape off teeth, are deposits of fat, cholesterol, and calcium that gradually accumulate on arteries of all diameters. The plaque causes the artery to harden and dilate. High blood pressure is one consequence. Smooth-muscle cells cause other problems when they creep from the vessel wall into the plaque, where they proliferate and calcify. (I describe the amazing way cells move around, a.k.a. migrate, in chapter 10.) Eventually, the growing plaque causes the artery to clog or rupture, wreaking havoc—stroke, heart attack, blindness, black toes—depending on its location. A healthy lifestyle helps slow the process, as do a variety of medications. To avoid an impending blockage or to treat a recent one, doctors have multiple means to approach the problem mechanically. The direct approach involves surgically removing the blockage by opening the artery longitudinally, scraping out as much of the plaque as possible, and then closing the vessel, often with a patch to further widen the vessel. This procedure, an endarterectomy, works only in larger vessels (for example, femoral artery in the groin, carotid artery in the neck) and where the major portion of the plaque affects only a short and accessible segment of the artery.

Suitable for smaller vessels, especially on the heart's surface, a less invasive technique (angioplasty) entails a cardiologist accessing the arterial system through a large artery, typically in the groin, sometimes in the arm. The doctors advance a catheter into the diseased coronary artery by watching a real-time motion X-ray image on a video screen. Then they thread the catheter into the narrowed area and repeatedly inflate a tiny balloon to stretch the artery back toward its original diameter. To preserve the expansion, they "stent" the vessel open by slipping an expandable tube of meshed-metal scaffolding into the diseased area. (Little-known fact: these expanders that hold the artery open are called stents after nineteenth-century British dentist Charles Stent, who devised a special wound dressing that would remain in place inside the mouth. It is likely that as long as ASCVD persists, his name will remain in use.)

The next way of approaching dangerously large plaques, especially when they involve a long segment of an artery, is to bypass the offending section, thereby rerouting the passage of blood around an artery's restricting plaques. Often the bypass material is a segment of one of the patient's veins, which is readily available, entirely biocompatible, and expendable because the body's venous system has marked redundancy. When a sufficiently long or large-diameter vein is not available, woven-fabric tubular grafts of Dacron® or Teflon® have proven useful.

Cardiac surgeons also use vein grafts to bypass obstructed sections in coronary arteries. The expression "a quadruple bypass" means that diseased sections in four coronary arteries were addressed and implies more extensive disease than what might be managed by a single, double, or triple bypass. The procedure is known formally as a coronary artery bypass graft (CABG), or a "cabbage." Surgeons originally used the patient's own veins as the bypass conduits, but when veins transferred in this way experience the unaccustomed arterial pressures, they tend to develop atherosclerotic plaques of their own, leading to recurrent blockage. Grafted arteries are less likely to develop this problem, so an artery in the forearm, the one where you can feel your pulse at the wrist, is a preferred donor, but only if the hand is adequately supplied by alternate sources, which it usually is.

Neither veins nor fabric grafts fully substitute for a healthy smooth muscle–endowed artery with its resiliency and autonomic control. Under investigation, another means of artery replacement in ASCVD is the use of bioengineered arteries. In the laboratory a technician seeds smooth-muscle cells onto a biodegradable tubular meshwork of the desired length and diameter. This composite is then incubated in nutrient broth. Over weeks, the cells multiply and produce a tough, fibrous cellularized tube, and simultaneously the meshwork dissolves. Using various chemicals, the technician then removes all the living cells and their proteins. This step prevents the graft from causing an immune response when this fibrous, biologically constructed tube jumps into action.

Another approach to treating ASCVD is useful mostly for the smallest vessels, such as the arteries supplying the toes. At roughly a sixteenth of an inch in diameter, these arteries are too small for endarterectomy, dilation and stenting, or bypass grafting, so doctors turn to interrupting the sympathetic nerve supply that tends to keep arteries in the distant parts of the limbs constricted. One method is to inject a paralyzing agent along the course of the nerve. This can provide vasodilation for months until the medication's effect wears off. The other method entails endoscopic surgery, which physically disrupts the sympathetic nerves to the affected areas for all time.

One final note on arterial smooth muscle—it sometimes serves a lifesaving function when an artery suffers a laceration and blood spurts out in time with the heartbeat. It is intuitive to apply pressure to the wound to minimize blood loss, and that certainly helps, but the smooth muscle in the arterial wall goes into overdrive to stanch the flow. The vessel's circumferentially oriented smooth-muscle fibers contract to narrow the opening in the arterial stump. The artery's natural elasticity also pulls the bleeding ends back into the tissues surrounding the cut where stabilizing clots can more easily form.

Apart from when a blood vessel is cut, the circulatory system is our one set of tubes that has no opening to the outside world. The other body systems endowed with smooth-muscle tubes (respiratory, digestive, urogenital) have at least one natural external opening. Each of these systems' smooth-muscle arrangement and function are unique and can face maladies both common and bizarre.

Gastrointestinal Problems

The digestive system is a long smooth-muscle tube with openings top and bottom. Food in the tube first encounters smooth muscles in the throat. They involuntarily contract to advance each swallow of dinner into the esophagus. The esophagus has sphincters—ringlike muscular closures—at each end.

Based on the results of microscopic scrutiny in the laboratory, the upper sphincter is classified as a voluntary muscle, although I don't know if, how, and when I contract mine. The lower one is where the esophagus empties into the stomach. It works involuntarily and closes off the esophagus entirely unless a bite of dinner is passing by. Both swallowed solids and liquids make their way down the esophagus past one sphincter and then the other via peristalsis, which means that one can effectively advance food into the stomach while lying down or, purportedly, even when doing a handstand.

As in the esophagus, sphincters close the stomach portion of the digestive tube at each end. So contained, food gets thoroughly stirred by the stomach's three layers of smooth muscle before the gastroduodenal sphincter opens, allowing advancement of stomach contents into the small bowel. About 22 feet further on, the food, now completely broken down and its nutrients absorbed into the bloodstream, encounters the next involuntary sphincter, which is the gatekeeper to the large intestine. This portion of the digestive tract is a 5-foot-long smooth-muscle tube that absorbs water left over from all the preceding action.

At the large intestine's terminus, two adjacent sphincters close off the digestive tube from the outside world. The first is involuntary, the second voluntary. The redundancy suggests that they perform an important function. In 1960, Dr. Walter C. Bornemeier described it this way:

> They say man has succeeded where the animals fail because of the clever use of the hand, yet when compared to the hands, the

> sphincter ani [anal sphincter] is far superior. If you place into your cupped hands a mixture of fluid, solid, and gas, and then through an opening at the bottom, try to let only the gas escape, you will fail. Yet the sphincter ani can do it. The sphincter apparently can differentiate between solid, fluid, and gas. It apparently can tell whether its owner is alone or with someone, whether standing up or sitting down, whether its owner has his pants on or off. No other muscle in the body is such a protector of the dignity of man, yet so ready to come to his relief. A muscle like this is worth protecting.

If you agree that farting is a miraculous feat, let's also consider the act of burping. A common malady of gastrointestinal smooth muscle occurs when the involuntary smooth-muscle sphincter between the esophagus and the stomach does not stay closed between swallows. In that case, gastric contents, which are highly acidic, can rise into the esophagus. This is the source of heartburn, also informally known as acid indigestion and formally as gastroesophageal reflux disease (GERD). I think it's safe to say we've all had at least momentary experience with that.

Because the sphincter does not make a complete seal, our symptoms may worsen when we are bending over or lying flat. Furthermore, tight clothing can pressurize the stomach and promote

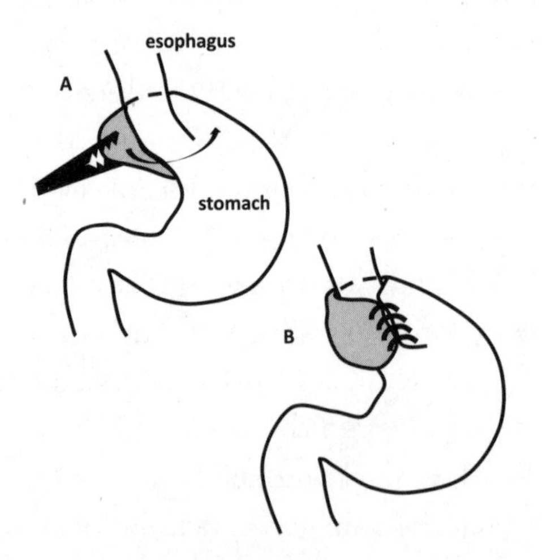

One method of treating esophageal reflux is to wrap an upper portion of the stomach around the esophagus to strengthen the gastroesophageal sphincter.

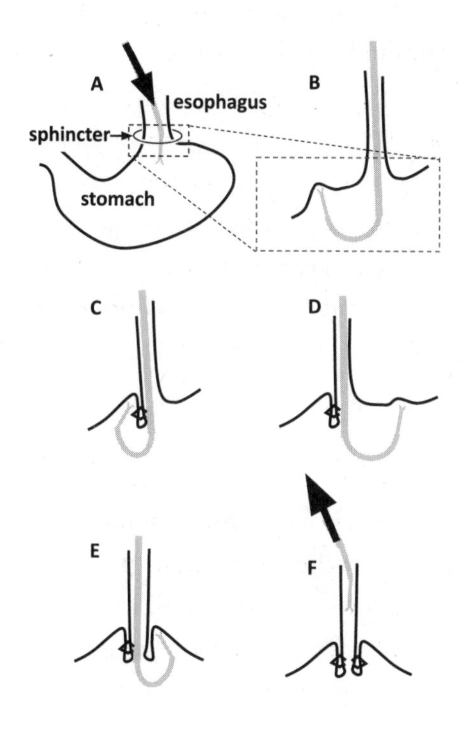

Technique of transoral incisionless lower esophageal sphincter augmentation. (A) To tighten the sphincter, a flexible instrument is inserted through the mouth into the stomach. (B) The instrument grasps the stomach wall and folds it back on itself, (C) secures it with a staple, and then (D) grasps the opposite wall and folds it similarly. (E) On completion, the muscular junction between the esophagus and the stomach is thickened and capable of resisting reflux of gastric contents upward.

backflow. Eating small meals and eating them slowly reduce volume overload. And some over-the-counter remedies are pretty effective at diminishing gastric acidity and hastening the transit of stomach contents past the next sphincter and into the small intestine, which manages gastric acid just fine.

When such remedies fail, several novel surgical procedures can substitute for the lower esophageal sphincter. The classic procedure wraps a portion of the stomach around the esophagus for added pressure. Remarkably, doctors can now perform this procedure either endoscopically through several small incisions in the abdominal wall or without any incisions at all. Along with the others, the latter procedure also requires anesthesia (thank goodness) and entails sliding a complicated flexible instrument over an endoscope and passing both into the stomach through the mouth.

These techniques use the patient's own smooth muscle for sphinc-

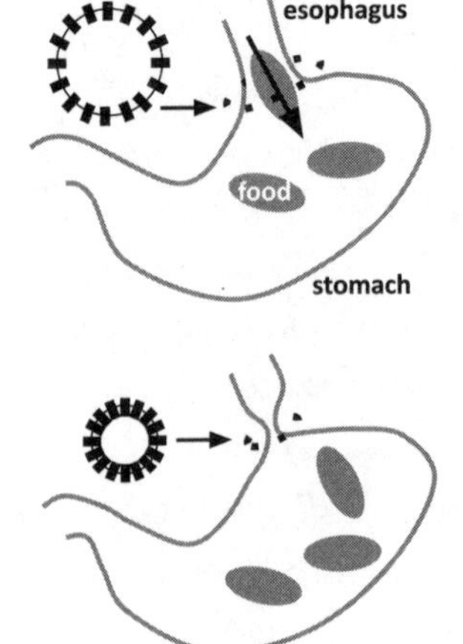

Placed around the gastroesophageal junction, a ring of magnetic beads, which are naturally attracted to one another, momentarily enlarges to allow food to pass from the esophagus into the stomach. The ring then tightens to prevent stomach contents from reentering the esophagus.

ter supplementation. Doctors can also supplement the strength of either the gastroesophageal sphincter or the anal sphincters by circling the deficient sphincter with a ring of magnetic beads whose attraction to one another holds the sphincter closed. Increased pressure from inside the tube breaks the magnetic beads' mutual attraction to allow the ring to open temporarily to let matter pass through.

The small bowel (small intestine) is where absorption of digested nutrients occurs. At times, an infant is born with an extremely short small bowel or an adult acquires a similar deficiency from a catastrophic blood clot or infection. The result is malnutrition. One novel remedy is to slow the passage of food and water through the available short segment to provide more time for absorption to occur. This entails surgically reversing a several-inches-long segment

When the small bowel is too short to allow for absorption of nutrients, the transit speed of digested food can be slowed by surgically reversing a segment of the bowel. Reverse peristalsis in this segment slows the movement of food and allows more time for nutrients to be absorbed.

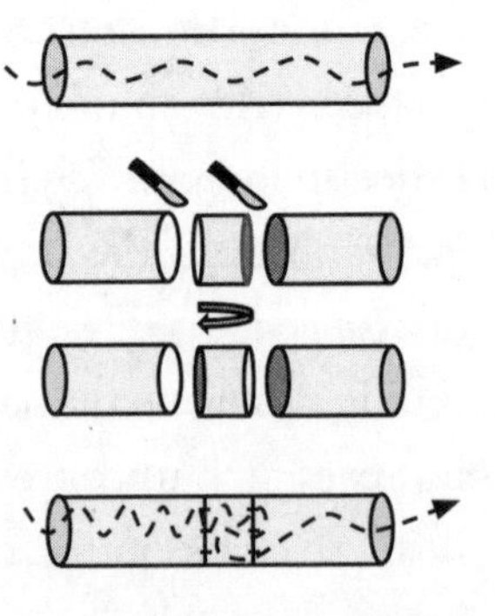

of the existing small bowel, which puts brakes on the usual direction of peristaltic action and aids digestion.

If surgery is not possible or proves ineffective, the person must receive nourishment multiple times daily intravenously. Over time, this proves to be neither entirely feasible nor safe, which leads to the consideration of a small bowel transplant. Living donors can easily relinquish 6 feet of their small bowel without experiencing any digestive consequences, and a recently deceased donor can relinquish the whole thing, with the liver attached, if necessary. Blood vessels are reconnected in the recipient to reestablish circulation, but the autonomic nerve fibers supplying the transplant are microscopic, diffuse, and defy connection. No problem. After several days, vital peristalsis resumes spontaneously, courtesy of the small bowel's neural network.

Urinary Upsets

Swallowed water is absorbed into the blood from the large intestine and then is filtered out by the kidneys, which also scavenge metabolic wastes from all over the body. Liquid wastes and excess water are then sent packing through the ureters, bladder, and urethra, all of which are well endowed with two or three laminations of variously oriented smooth-muscle fibers in their walls. Similar to the terminal end of the digestive system, the outlet of the urinary system between bladder and urethra is also controlled by two sphincters, one involuntary and one voluntary. They do not have, nor do they need, the discriminatory capabilities of the anal sphincters. In men, the two sphincters are separated by the prostate gland, which unfortunately can also, and undesirably, retard flow when enlarged. If medications are not sufficient to reduce prostate size, surgery is often needed.

The problems that occur far more often than retarded urine flow, however, are uncontrollable flow and fear of sudden involuntary flow. The American Urological Association indicates that such problems affect one-fourth to one-third of Americans, but firm statistics are elusive because many who have the problem consider incontinence a normal part of aging and may not seek help. Relief, however, is available.

Urinary incontinence is not a disease; rather it is a symptom of many different conditions, which vary between men and women. One of the most common forms is stress incontinence, which frequently occurs in older women. Childbirth markedly stretches the muscles across the base of the pelvis, and they may never fully regain their pre-pregnancy strength. These are skeletal muscles, under voluntary control, not the involuntarily controlled smooth muscles, which are the focus of this chapter. Nonetheless, I describe the condition here since it involves muscles of both types. Activities such as bending over, coughing, or lifting apply pressure to the bladder. If the smooth-muscle sphincters cannot fully resist, urine squeaks out. Medications do not seem to help. In 1948, gynecologist Arnold Kegel described a series of exercises to strengthen the skeletal muscles in the pelvic floor including the voluntarily controlled sphincter, and physical therapy specialists can oversee conditioning exercises to improve muscle tone in this sensitive area.

The other common form of urinary incontinence is OAB, overactive bladder. Here the brain and the smooth muscle in the bladder communicate the message that urgent emptying is needed, even though the bladder is nowhere near full. This may require multiple nocturnal trips to the bathroom as well as constant daytime vigilance regarding the proximity of a relief station. Certain foods may be triggers, as is caffeine. Kegel exercises to strengthen the voluntarily controlled urinary sphincter, fluid restriction, and scheduled bathroom trips may help. Frequently beneficial are oral medications or botulinum toxin infusions into the bladder cavity; both treatments suppress the parasympathetic nerve impulses that cause the bladder smooth muscles to contract.

A third line of treatment demonstrates the complexity of the interplay between the various parts of the nervous system and muscles, both voluntary and involuntary. It involves electrically stimulating a nerve at the ankle. It is the one that activates the calf muscles and provides sensory awareness to the sole of the foot. Stimulating this nerve at the ankle can quell OAB symptoms, apparently because activating the nerve supplying the calf muscles (voluntary) suppresses the activity of the one supplying the bladder (involuntary). This is remarkable since the two nerves are as separate from each other as night is from day. This is one of many yet-to-be-explained enigmas about muscle.

In contrast to bladder overactivity, in complete spinal cord injuries the bladder is disconnected from its nerve supply, cannot contract, and is therefore incapable of willful emptying. Catheters for drainage are not a good long-term solution because they risk bladder and kidney infections. A novel alternative is using a skeletal muscle to substitute for the paralyzed bladder's smooth muscle. It works when the level of the spinal cord injury has left the abdominal wall muscles with normal innervation yet paralyzed the bladder muscles. The procedure involves splitting the abdominal wall's six-pack muscle in the midline of the belly, detaching half of it from the ribs, folding that half down, and wrapping it around the bladder. Then, voluntarily creating an isometric contraction in the abdomi-

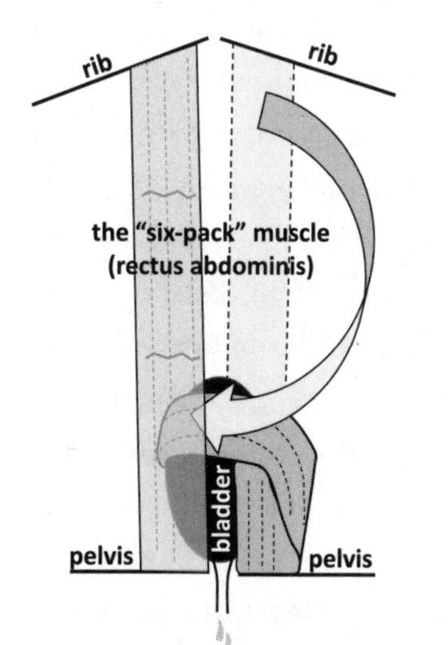

A paralyzed bladder associated with a spinal cord injury can be reactivated by wrapping it with a half of the abdominal wall "six-pack" muscle. The patient can voluntarily contract the transferred muscle to empty the bladder.

nal wall muscle, such as when preparing to cough, gives the bladder the needed squeeze.

Reproductive Ailments

When another smooth muscle, the uterus, maximally contracts, it can do so with great force and deliver a newborn baby. When the owner is not pregnant, uterine contractions periodically aid the egress of menstrual blood. In this instance, chemical messengers from the tissue lining the uterus induce the smooth muscle to contract frequently, forcefully, and irregularly. These contractions, along with reduced uterine blood flow and increased nerve sensitivity, can induce painful menstrual cramps.

Under other circumstances, rhythmic uterine contractility aids the upward passage of the sperm swarm racing to fertilize an egg. While the uterus is helping sperm move upward, peristalsis in its fallopian tubes moves an egg downward toward the oncoming swarm.

Benign smooth-muscle tumors in the uterus are commonly known as fibroids, and they affect as many as 80 percent of women, with some variation in noted incidence according to race. In face of their common prevalence, relatively little is known about the molecular mechanisms that regulate fibroids' appearance and growth. Even though they consist of smooth-muscle cells, fibroids are too disorganized to contract. A search for the cause of fibroids is confounded by the fact that multiple fibroids in one uterus can have different biological characteristics, including different growth patterns. When these smooth-muscle tumors remain small, they create no problem and may remain undetected. If they enlarge, they can first be detected during a pelvic examination and later by feeling them through the abdominal wall. Neglected, fibroids can become huge and mimic pregnancy.

Long before that, however, the combination of a fibroid's size and location on the uterus can cause abnormal bleeding, pelvic pain and pressure, back pain, urinary frequency, constipation, or infertility, singularly or in combination. Gynecologists direct medical treatment at reducing symptoms or shrinking the size of the fibroid. Depending on the size and location of the tumor and the woman's reproductive status, the doctors might recommend removal of just the fibroid or the entire uterus (hysterectomy). One radiologic intervention is to thread a catheter into the uterine artery, block it, and starve the fibroid of its blood supply. Another is to insert a probe under magnetic resonance imaging control and destroy the tumor with ultrasonic energy.

Fibroids can cause uterine factor infertility (UFI), which affects about one in every 500 women. This umbrella term covers conditions in which the uterus is developmentally absent or markedly deformed as well as when the uterus is diseased or has been removed. Traditionally, women with UFI remained childless or had a surrogate mother carry their embryo following in vitro fertilization. But thanks to advances regarding uterine transplants, in 2014 a woman in Sweden with UFI became the first mother to successfully nurture a fetus and deliver a baby using someone else's uterus. Now worldwide, over 100 uterine transplants have been performed, and about 20 live births have resulted.

To start the process, eggs from the woman with UFI are harvested, fertilized, and frozen. An altruistic woman who is finished with childbearing may donate her uterus, or it may come from someone recently deceased. The autonomic nerves controlling uterine smooth-muscle contraction are quite small and difficult to find at surgery, so they are not reconnected at the time of transplantation. Months later, the transplant is deemed successful when the uterus has gone through several menstrual cycles, which are cramp-free since the uterine smooth muscle now has no nerve supply. Then one fertilized egg is thawed and implanted.

As the pregnancy nears completion, labor contractions will not occur since the uterus has no nerve supply, so the baby is delivered by cesarean section. After one or two pregnancies, the mother receives a hysterectomy and can thereafter stop her antirejection medications. Obviously, this is a complicated procedure and requires the close attention of specialists, including those practicing high-risk obstetrics, radiology, pathology, neonatology, transplant immunology, ethics, psychiatry, and social work. It is clearly experimental but exemplifies the possibilities for smooth-muscle transplants.

Or maybe a donor uterus is not needed at all. That is the case for at least some laboratory rabbits at Wake Forest University. Researchers there recently reported that they had created uteruses via tissue engineering. They seeded endometrial lining cells and smooth-muscle cells onto a biodegradable meshwork scaffold and grew and divided them many times over in the laboratory. They sewed the fabricated uteruses into rabbits, which subsequently became pregnant and eventually delivered normal offspring by cesarean section. C-section was necessary because, just like a transplanted uterus, a fabricated uterus has no nerves supplying the smooth muscle, so the uterus cannot contract.

But there's more—currently hypothetical but worthy of consideration. Perhaps 1 percent of people considered to have male characteristics at birth have gender dysphoria and therefore identify as women. For some, their dysphoria so profoundly affects them that they undergo gender-affirming surgery. For some, receiving a uterus may be a requisite of gender affirmation. Again, the donor could be either recently deceased or even someone who is undergoing a hysterectomy as part of female-to-male reassignment surgery. To date, no such transplants have been performed, but they may be on the horizon.

Respiratory Issues

If a uterus transplant is not enough to take your breath away, consider the roles of smooth muscle in the respiratory system along with several related disease processes. The trachea (windpipe) begins just below the vocal cords, and after descending about 4 inches, it branches into the left and right bronchial tubes, which then split repeatedly inside the lungs into smaller and smaller airways (bronchioles). The trachea contains a series of springy, cartilaginous, horseshoe-shaped rings that hold it open regardless of the pressure inside the chest. Spanning each ring's opening is a strand of smooth muscle. During a cough, these smooth muscles involuntarily contract, which narrows the airway. This causes the expired air to accelerate and carry with it any accumulated mucus, dust, or other irritant.

Severe tracheal damage, for instance, from prolonged intubation and ventilator support for severe COVID-19 infections, is hard to correct. Tissue-engineered tracheas have failed, and transplants have until recently been considered impossible because the trachea is supplied by a series of tiny blood vessels that are too small to reconnect. In 2021, however, surgeons at Mount Sinai Hospital in New York got around this problem by transplanting the donor's thyroid gland and a section of esophagus along with the trachea as a single "block." (The arteries leading to and the veins leading from the esophagus and thyroid are suitably sized for repair, and tiny branches continue through these organs to supply the trachea.) The patient began breathing normally and proceeded to repopulate the transplanted lining cells with her own. The transplanted thyroid gland and section of esophagus served no purpose other than supplying blood vessels that were large enough to suture—but even so, they are just about a sixteenth of an inch in diameter.

Smooth-muscle fibers encircle the bronchi and can pose a different and far more common vexation. They constrict in response to various stressors, thereby narrowing the airways and impeding efficient air exchange—causing what's known as asthma. It is hard to understand how bronchial constriction might have ever served some useful purpose, and people with asthma have reason to damn it. (Bronchial smooth muscle has been called the appendix of the respiratory system since it provides no apparent benefit and just causes trouble. The analogy is only partly true, however, because a misbehaving appendix can be removed without consequence while a constricted bronchus cannot.) Management of asthma centers on inhaled medications—corticosteroids to reduce inflammation and chemical blockers that prevent the autonomic nerve signals from reaching the smooth muscles. When medications cannot adequately control asthma, it is possible to insert an endoscope into the bronchi and heat the smooth muscle and bronchial nerve endings to permanently destroy them.

◆◆◆◆

IN GENERAL, SMOOTH muscle serves us well. It typically works without our awareness while taking responsibility for functions such as accommodating eyesight, maintaining a steady body temperature, and transporting semisolids, liquids, and gases into and out of the body while serving as gatekeepers at critical junctures. We should appreciate smooth muscles and be happy that they work involuntarily. Consciously monitoring all of them would be a burden. The same is true for the next type of muscle, which is reserved for a single organ.

Chapter 5

CARDIAC MUSCLE

A THREE-WEEK-OLD HUMAN EMBRYO IS ABOUT THE SIZE OF A peppercorn. It already has a functioning heart, which can continue purposeful contractions for as long as 120 years.

Once the heart grows to its adult size, there is precious little replacement of its muscle cells. At age 25, approximately 1 percent of them are replaced annually, and by age 75, the turnover rate is half that. For comparison, cells exposed to harsh and changing environments get replaced far more often. For example, cells lining the intestinal tract have a longevity of about four days, and those constituting the skin live two to three weeks. Brain cells, however, have cardiac muscle beat for durability. They are never replaced.

Depending on the size of the individual, the heart weighs the same as two to three sticks of butter. It is about the size of a fist, but the size changes according to whether the main parts (ventricles) are contracted or relaxed, a change that occurs between 40 and 200 times a minute depending on the owner's age, level of conditioning, and intensity of activity. Each ventricular contraction sends about 2.5 ounces of blood on its way. That is enough to fill an ordinary aluminum soda can in about four seconds or a small tanker truck (1,500–2,000 gallons) in a day.

When blood first enters the heart, it gets pumped to the lungs, where it exchanges its carbon dioxide for recently inhaled oxygen. Then the blood returns to the other side of the heart to start on its oxygen-delivery journey throughout the body. Leaving your heart, the blood speeds along at about a foot per second, so four seconds after getting squeezed out, it is already down past your knee. Once it has passed through capillaries, where it relinquishes its oxygen and other nutrients and picks up carbon dioxide and other metabolic waste products, it passes through progressively larger veins before reaching the heart once again. In fact, almost all our tissues receive their nourishment in this manner. (The exceptions are cartilage, which surfaces our joints, and the cornea. They receive nutrients by soaking up fluid from adjacent tissues, which themselves are nourished by capillaries.)

It is hard to understand the heart's arrangement and function by just looking at even the most lifelike drawing because its four chambers and intake and exhaust ports fold over and around one another in a complex three-dimensional arrangement. This allows the heart to fit compactly into the chest and to optimize pumping efficiency, but it makes it hard to depict graphically. Conceptually, the human heart is a linear sequence of four contractile chambers with the lungs

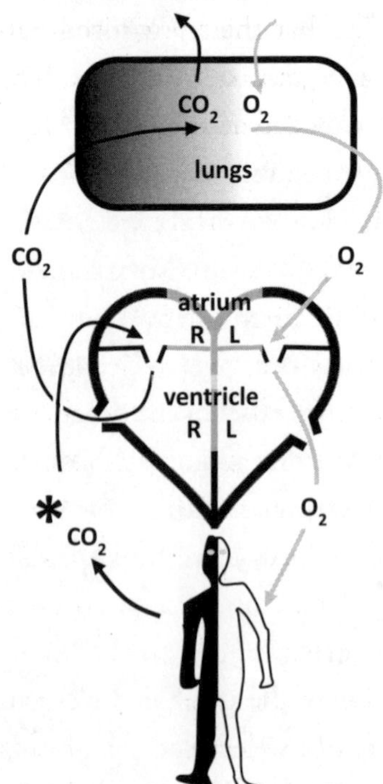

Starting at the asterisk, blood carrying carbon dioxide (CO_2) from the body enters the right atrium and is pumped to the lungs by the right ventricle. In the lungs, CO_2 is exchanged for oxygen (O_2). The blood then returns to the heart's left atrium. Left ventricular contraction circulates the oxygen-rich blood throughout the body.

halfway along—a small chamber (right atrium), a large one (right ventricle), lungs, another small chamber (left atrium), and another large one (left ventricle). Four one-way valves keep the blood moving in the proper direction with each cardiac contraction, which is far more forceful and frequent than the peristalsis seen in the body's smooth-muscle tubes.

And just as intestinal and fallopian tube peristalsis is not entirely efficient, neither are the cardiac contractions, mostly because the chambers cannot completely collapse to squeeze every drop of blood along. In fact, a properly functioning heart will eject only 50–70 percent of the blood contained within it during one contraction. A mechanical engineer might not find that percentage very impressive for a pump, but considering the heart's durability and longevity, I think it is worthy of great respect.

To account for such superlatives, the actin/myosin units in cardiac muscle fibers are arranged in a far more orderly fashion than are the units in smooth muscle. In fact, the units line up so perfectly, both side to side and end to end, that their organized pattern is visible under magnification. Antonie van Leeuwenhoek, who invented the first microscope, noted this banded or striped nature of "striated" muscle in 1682. He was likely examining skeletal muscle, which, along with cardiac muscle, is striated. Now it should be clear why smooth muscle is called smooth—it is not as highly organized as cardiac and skeletal muscle and does not have striations.

So, since they are both highly organized and striated, what then is the difference between cardiac and skeletal muscle? The heart contracts spontaneously on its own, with some oversight provided by the autonomic nervous system's two competing/cooperating parts as well as by circulating hormones. By contrast, skeletal muscles are mostly under voluntary control, which stems from a conscious demand generated in the brain's motor cortex.

Remember that the sympathetic portion of the autonomic nervous system controls fight-or-flight functions, so its electrical signals will cause the heart to beat faster and more forcefully. The heart also picks

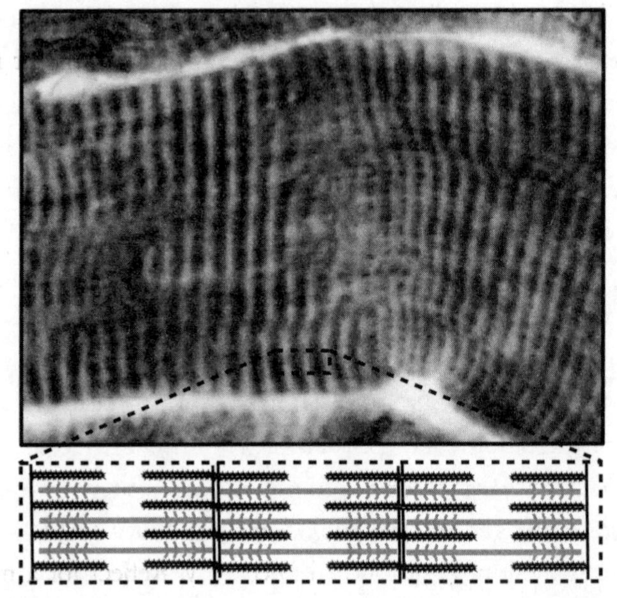

Under high-magnification light microscopy, both cardiac and skeletal muscle fibrils demonstrate a striped (striated) pattern. In the photomicrograph, six striated muscle fibers are oriented horizontally. In the enlarged schematic, the repeating actin/myosin units arranged in a very orderly fashion account for the striations.

up the pace acutely in response to circulating adrenaline released in moments of stress by the adrenal glands, sitting atop the kidneys. Thyroid hormone stimulates the heart rate over the long haul. Conversely, parasympathetic cardiac stimulation causes the pulse to slow.

Left alone, however, the heart typically does just fine without autonomic nervous system control. Recall two points described previously. First, the heart begins working when the embryo is minuscule and when there is only a rudimentary nervous system at best and no adrenal glands at all. Second, the small intestine has its own neural network and can continue rhythmically contracting even when completely separated from all nerve control. The heart is similar, but instead of a diffuse neural network it has three "in-house" pacemakers, one at the top of the right atrium and two others deeper within.

These pacemaker cells are specialized cardiac muscle cells that spontaneously generate electrical signals, which then, under normal circumstances, quickly spread in an orderly sequence throughout the heart. The generation and transmission of each electrical signal causes each cell's membrane to momentarily lose its electrical charge. This triggers atom-wide channels to open, and calcium ions flow into the cell to activate muscle contraction. Once the impulse has passed, each cell recovers its resting state. In less than a second (far sooner in hummingbirds), the system resets and is ready to conduct the next electrical impulse. It is good, however, that the impulse does not travel immediately and uniformly through the entire heart. Rather, the atria contract first, advancing blood into the ventricles; then a fraction of a second later, the right and left ventricles contract, expelling their contents toward the lungs and body, respectively. Reflect for a moment that this self-generating sequence of vital electrical-to-mechanical transformations can occur, sometimes flawlessly, for about three billion lifetime contractions in both hummingbirds (1,260 times per minute for 5 years) and in humans (70 times per minute for 100 years).

The specialized cardiac muscle pacemaker that sits atop the right atrium regulates the heart's normal rate of contraction, which is usually between 60 and 100 beats per minute and slower in highly conditioned athletes at rest. The second pacemaker is subservient to the first one, unless the first one goes haywire. Then the second one takes over and generates contractions at 40 to 60 beats per minute, perhaps leaving the owner short of breath but certainly allowing sufficient time to have the heart checked out and the pulse normalized. Remarkably, a third pacemaker takes over if the first two fail, but it generates contractions at only 20 to 30 beats per minute, generally too slow to maintain an adequate blood pressure.

Voluntary Control

I previously mentioned that cardiac muscle is beyond our control—for the most part. There are several interesting exceptions. One you can demonstrate to yourself right now. Feel your pulse at your wrist or put the tip of your index finger at the top of your neck directly below your ear to feel the pulse in your carotid artery. (That is how I'd confirm that I was still alive during boring classes. It looked like I was just resting my chin on my hand.) Feel your pulse for a few beats, then exhale and hold your breath while contracting your chest and abdominal wall muscles, as if you're having a bowel movement. At first, this squeezes blood that is already in the chest into the heart, making it work harder. Then the pressure prevents more blood from entering the chest, so the heart has less to do. Your pulse will slow. Let go and take a deep breath. After a momentary lapse, as blood reenters the chest, then the lungs, then the heart, your pulse will pick up.

Other ways to take charge of your heart rate either demand great practice and concentration (yoga) or require putting your face in water, the colder the better. All vertebrates, terrestrial as well as aquatic, manifest the dive reflex, which includes parasympathetically mediated slowing of the heart and sympathetically mediated constriction of arteries supplying the limbs. Both functions ensure that vital organs remain adequately supplied with oxygen. The champion breath holder is a Cuvier's beaked whale, recorded to stay underwater for almost four hours. The scientists were not able to monitor its pulse; but others have monitored it in blue whales, whose pulse on the ocean surface is about thirty beats per minute and drops to as low as four during deep dives.

Ever-curious and enterprising researchers have used laboratory rats as experimental models. The researchers trained rats to voluntarily swim a maze that first was on the surface of an aquarium and then submerged. Lead author W. Michael Panneton reports, "When placed in the start area, the rats voluntarily initiated their own dives to reach the finish area. The rats appeared unstressed by either the training protocol or the exposure to water. No external reward was used during the training protocol." Their heart rates dropped from about 450 to about 100 beats per minute while diving. Thanks, rats, for confirming that under certain circumstances, vertebrates do have some control over ordinarily involuntary muscles.

For further confirmation, I tried it myself by setting a timer, counting my pulse, and plunging my face in a bowl of ice water. It proved to be one of those scientific studies with a negative result that will never reach publication—the trial was too painful for me to remain immersed for more than 15 seconds, during which time I could not focus on taking my pulse, if I had one. With a waterproof watch or timer, you could test your dive reflex in a pool. The reflex seems to be triggered by immersing your face and getting some water in your nose.

Remote Monitoring

Monitoring the pulse is one way to study the heart; recording its electrical activity is another. An electrocardiogram captures and records the depolarization and repolarization of atrial and ventricular muscle cells. (Logic would suggest the word electrocardiogram should have the acronym ECG, but it is often called an EKG. Why? Because its name is derived from the Greek word *kardia*, meaning heart.)

What do the squiggly ripples and spikes on an ECG tracing indicate? The first blip signifies the atria depolarizing electrically and contracting mechanically. The flat interval that follows is the delay between atrial and ventricular contraction that allows the ventricles to fill. Then the sharp spikes indicate the ventricles contracting, and the trailing blip indicates the ventricles repolarizing. (Any evidence on the tracing of the atria's repolarization is masked by the ventricles' stronger electrical activity occurring at the same time.)

At times, the generation and conduction of electrical signals among the cardiac muscle cells goes haywire and results in irregular, inefficient heartbeats—arrhythmias. The most common one, atrial fibrillation, affects 2 to 3 percent of Americans and North-

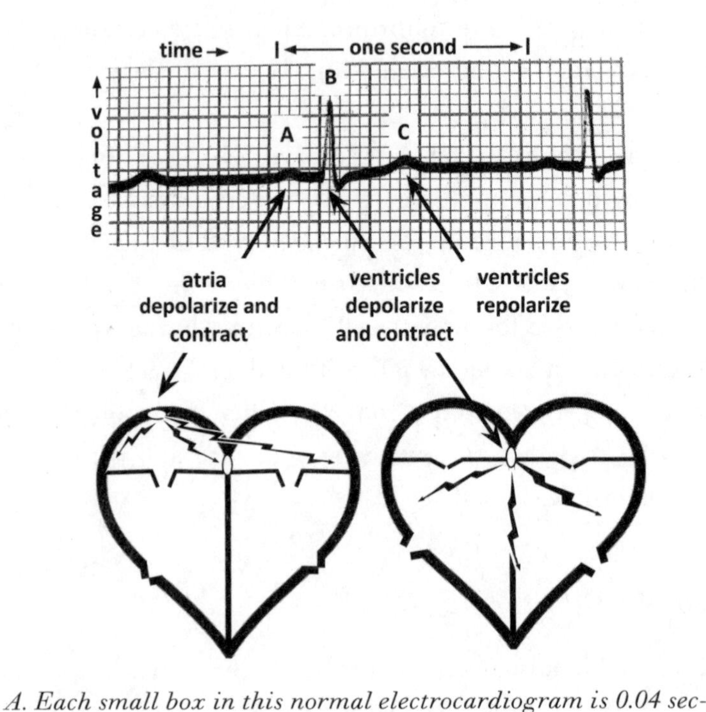

A. Each small box in this normal electrocardiogram is 0.04 seconds. The tracing shows a heart rate of 60 beats per minute.
B. An electrical impulse originates in the primary cardiac pacemaker in the right atrium, spreads through both atria, and causes them to reverse their electrical charge and contract.
C. The impulse reaches the secondary pacemaker 0.16 seconds later. It sends an electrical impulse to the ventricles, causing them to depolarize and contract. After another 0.35 seconds, the ventricles repolarize to regain their resting electrical charge. The heart muscle is then prepared to contract again. (The repolarization of the atria is masked on the tracing by the ventricles' stronger depolarization signal, which occurs simultaneously.)

ern Europeans. It is less common before age 50, but for people in their eighties, one in eight are affected. In atrial fibrillation, rogue cardiac muscle cells override the normal pacemaker's messages with their own uncoordinated electrical impulses, which cause the atrial cells to contract irregularly (fibrillate) at over 600 beats per minute whether or not they have had time to fill with

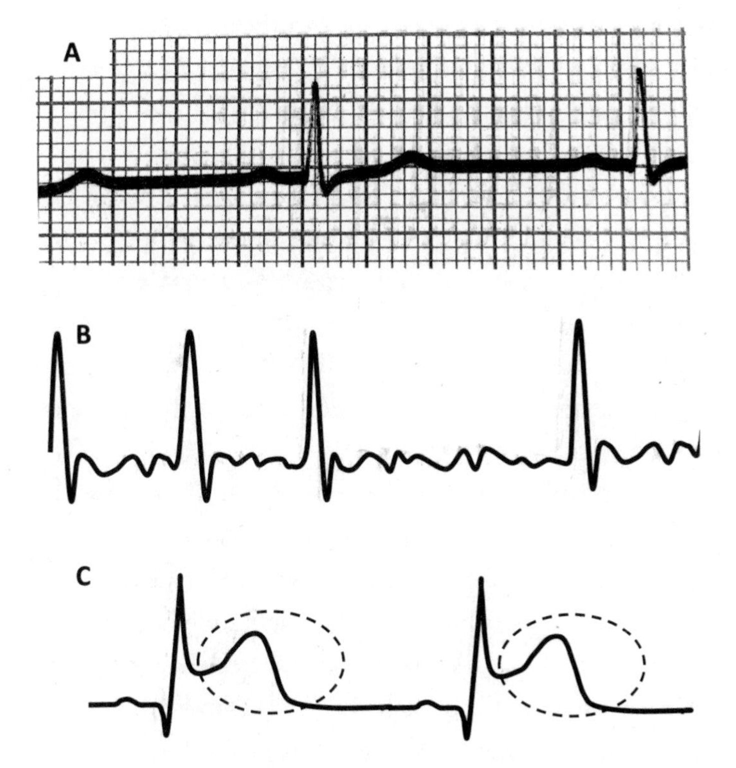

A. This normal electrocardiogram shows two heartbeats, with the large spikes indicating ventricular contractions.
B. Characteristic of atrial fibrillation, the large spikes indicate ventricular contractions, which are occurring at irregular intervals. The intervening electrical impulses are chaotic.
C. This tracing indicates an acute myocardial infarction. A marked voltage elevation is present at the completion of ventricular contraction and throughout ventricular repolarization (dotted circles).

blood following the previous contraction. The ventricles are similarly affected and contract at a slower, but also abnormally fast rate, which causes the pulse to be irregular and rapid.

Initially, episodes of atrial fibrillation may be brief and then gradually become longer and eventually continuous. Perhaps a third of patients with atrial fibrillation are unaware of their condition. Others may note skipped beats, exercise intolerance, and even chest pain if the heart itself is not receiving its necessary oxygen supply.

Atrial fibrillation may occur in an otherwise normal heart but is often associated with high blood pressure, valvular or coronary artery heart disease, lung disease, sleep apnea, and obesity. Treatment starts with managing any underlying condition followed by medications directed at either slowing the heart rate into the normal range or reestablishing normal rhythm. The latter sometimes requires cardioversion, which is a direct-current electrical shock through the chest wall (while the patient is sedated!) to restore control to the primary pacemaker cells. In stubborn cases, catheter ablation or surgery to inactivate the rogue atrial cardiac muscle cells is required to restore normalcy.

Even when atrial fibrillation is not causing heart-related symptoms, it poses danger to other organs, particularly the brain. Because the fibrillating left atrium is contracting irregularly and inefficiently, blood can swirl and form clots rather than entering the ventricles. If fragments of the clot break free and enter the circulation, they can cause a stroke if they reach the brain. Strokes certainly occur in the absence of atrial fibrillation, but atrial fibrillation approximately doubles the risk overall and accounts for about a third of all strokes in the elderly. The prescription of blood thinners to keep blood in the left atrium from forming clots is standard therapy for most patients.

The ECG alerts the cardiologist not only to the health of the heart's electrical system but also to the size and location of its chambers, the presence of drug toxicity or potassium imbalance, and much more—all from minuscule electrical impulses generated by the heart's specialized myocardial cells, which transmit them intentionally across the heart but also coincidentally and ever so faintly onto the chest's surface and out into the limbs, where the standard leads for an ECG tracing are placed.

Who knows what was in the mind of Scotsman Alexander Muirhead in 1872 when he first recorded a human's cardiac electrical impulses by attaching wires to his subject's wrists and ankles? Muirhead was an electrical engineer and is best remembered for

A commercially available EKG machine in 1911. The image's caption states: "Photograph of a complete electrocardiograph, showing the manner in which the electrodes are attached to the patient, in this case the hands and one foot being immersed in jars of salt solution."

his development of wireless telegraphy and selling seminal patents to Guglielmo Marconi, who pioneered radio transmission. Muirhead studied just one person, but other investigators followed and experimented widely. One enterprising scientist used a toy train to steadily advance the photographic plate used to receive a tracing of the electrical discharges.

By 1903, Willem Einthoven, a Brit, had developed a quite sensitive device, but it had a problem. It weighed 600 pounds and required five operators. Eight years later, ECG machines were commercially available but still required a desk-sized console, and the patient had to immerse a hand and a foot in separate tubs of water. (I think I could have tolerated the water tubs if they'd let me play with the model railroad.) Einthoven continued improving his device and receives credit for developing the first truly practical

ECG machine. For his contributions, he received the Nobel Prize in Medicine in 1924.

We can get a good sense of our skeletal muscles because they reside just below the skin, where they are available for inspection, palpation, and strength determinations. By comparison, the heart hides away inside the chest, so the means of evaluating it must be indirect. Also, it is best not to stop it just to have a good look. Blood pressure and pulse are basic indicators, and an ECG is revealing in many ways.

Coronary Artery Disease

Imaging studies began with Wilhelm Roentgen's discovery of X-rays in 1895 and have expanded into computed tomography, magnetic resonance imaging, and echocardiography (ultrasound). Injecting dyes that produce contrasting images formed by the heart muscle, valves, and circulating blood can enhance any of these imaging techniques. Echocardiography can examine the heart with placement of an ultrasound probe on the chest's surface and even more directly by passing the probe into the patient's mouth and halfway down the esophagus. Except perhaps for sword swallowers, patients undergoing transesophageal echocardiography require sedation.

The heart requires a lot of energy to constantly pump blood all through the body (with the noted exceptions of cornea and cartilage). To do so, it must supply itself with a constant flow of oxygen and nutrients, which it does via a system of arteries on its surface—the coronary arteries. Unfortunately, these arteries, especially the ones supplying chip-ingesting, couch-potato bodies, tend to narrow from the accumulation of atherosclerotic plaques, as I described in chapter 4. Should a plaque rupture and cause a coronary artery–blocking blood clot, the area of cardiac muscle dependent on that artery for nutrition dies. Call it what you will, a myocardial infarction, MI, or heart attack. Remember that healthy cardiac muscle cells typically survive for many decades.

That is the good news—they are durable. The bad news is that when starved of oxygen, they die. So does their owner when many cardiac muscle cells are simultaneously choked. If the owner survives the initial insult, the ECG pattern changes according to the size and location of the damage.

Also, the injured muscle releases two forms of an enzyme, troponin, which are unique to cardiac muscle. Hence ECG changes and elevated levels of these two forms of troponin indicate a myocardial infarction and differentiate it from other sources of chest pain—indigestion, for instance—whose symptoms may be similar. Damaged cardiac muscle is incapable of regenerating and turns to scar. The size and location of the scarred area variably affects the heart's capacity to pump efficiently. High blood pressure and physical exertion stress the remaining functional portions of the heart and put them at risk as well.

Medications may help keep the coronary arteries open and blood flowing. Angioplasty and stenting are also commonly used interventions, along with coronary artery bypass grafting. Investigational approaches are undergoing clinical trials. These include bioengineered patches that consist of either a natural or synthetic meshwork onto which muscle cells or muscle-precursor stem cells are embedded and cultured. After multiple rounds of cell replication in the laboratory, a surgeon attaches the composite patch to the surface of the heart. The transplanted cells heal to healthy cardiac muscle cells on the periphery of the scarred area and overgrow the scar, reducing its adverse effect on the heart's electrical and mechanical activity.

Coronary heart disease accounts for the death of about one-third of people in developed countries who are over 35 years old. The cardiac muscle and its owner are the victims, whereas the perpetrators are the clogged arteries, and I have previously described the role that smooth muscle plays in that tragedy. Typically, atherosclerotic cardiovascular disease provides warnings (high blood pressure, angina, shortness of breath) before it kills.

Cardiac Muscle Maladies

A warning is not in store with some other conditions that affect cardiac muscle directly even when the coronary arteries are perfectly fine. Hypertrophic cardiomyopathy, for example, is much less common than coronary artery disease, affecting about one in 500 individuals, but it can cause sudden death even for an apparently fit young athlete. It is the most common genetic cardiovascular disorder and is often an inherited condition, so somebody with a blood relative who suddenly dropped dead for no apparent reason should be particularly wary. A mutation in at least one of nine muscle genes, usually myosin or a myosin-modulating protein, accounts for overgrowth of cardiac muscle, particularly in the wall between the left and right ventricles. This thickening reduces the size of the left ventricle and the efficiency with which it fills and empties, thus putting the heart on overload, particularly under conditions of intense physical activity. Genetic testing is feasible but not practical because of the number of possible contributing mutations. When suspected, cardiac MRI, echocardiography, and ECG tests will lead to the diagnosis.

Doctors advise symptomatic individuals to avoid strenuous athletics, since the condition is a deadly time bomb. For those with symptoms, which include shortness of breath, exertional chest pain, light-headedness, and fainting, the same types of medications used for coronary artery disease may be useful. When activity restriction and medications cannot control symptoms, the best treatment is surgically removing a wedge from the hypertrophied muscular wall separating the ventricles and by improving the function of a valve.

An alternative to that drastic-sounding treatment of open-heart surgery is the creation of a "controlled heart attack" by inserting a catheter into the coronary artery supplying the interventricular wall and injecting alcohol, which destroys the overgrown muscle cells. This is obviously less invasive, but it cannot be as carefully tailored as the open-heart procedure and is therefore generally reserved for patients who are high surgical risks due to conditions such as frailty, advanced age, and severe lung disease.

Coronary artery disease and hypertrophic cardiomyopathy are only two of several ways that cardiac muscle can go bad. Consider myocarditis. The suffix *-itis* means inflammation, which is characterized by localized redness, swelling, heat, and possibly pain; it is the body's reaction to an injury or infection. Myocarditis can occur in conjunction with various viral infections, including COVID-19. Myocarditis has occurred in people who have been immunized via the Pfizer/BioNTech and Moderna vaccines, but it is seven times more likely to be seen in those actually infected with COVID-19.

Heart failure is the final common path of most heart disorders, whether the cause was coronary artery disease, hypertrophic or other types of cardiomyopathy, inflammation, valve problems, hypertension, or congenital malformation. Structurally and functionally, the heart loses its capacity to nourish the body's 30 trillion cells. Heart failure affects 2 percent of all adults and 6 to 10 percent of those over 65. About 35 percent of those with the condition will die within the first year after heart failure is recognized. This is because as the heart fails in some combination of rate, force, and volume of contraction, blood backs up and causes swollen legs, breathlessness, and excessive tiredness. Chest pain is always a feature. Cessation of smoking, weight management, and supervised exercise are first-line treatments, followed by medications, which vary according to the underlying cause.

More complicated treatments are available for such a common and potentially lethal condition as heart failure when medications fail to control it. An early surgical procedure that demonstrates the out-of-the-box thinking that surgeons around the world have applied to the ubiquitous problem of heart failure is noteworthy. It entailed converting a skeletal muscle to an auxiliary cardiac one. The latissimus dorsi is a large flat muscle that arises from the posterior trunk and attaches to the arm near the shoulder. It draws the overhead arm to the side, useful particularly when climbing. When somebody's heart is failing, circulation takes precedence over pull-ups, however, so the idea was to detach the muscle from the back while leaving it attached to the uppermost part of the arm. Then, by cutting away a section of the second rib, the surgeon rotated the muscle into the chest and wrapped it around the failing heart. An applied implanted pacemaker then stimulated the muscle to contract in time with the heart. It sounds heroic, but a desperate disease must have a desperate cure. Some years after its introduction, and with frankness rarely seen in medical publications, Dr. Carl Leier added this perspective: "In short, those who need it, don't survive it, and those who survive it, don't need it."

Conceptually appealing but not yet living up to its hype is implantation of stem cells into a failing heart. Once in place, either via coronary artery injection or direct injection into the heart itself, the cells could mature into functioning muscle cells and potentially replace failed native cells. To date, however, the intriguing possibility has been besmirched both by a number of research papers from a leading lab that were withdrawn for apparent fraud and by a rush to commercialization. At least 61 businesses in the United States were recently offering stem cell therapy for heart failure, although no treatments are FDA-approved. In 2019, Dr. Gregory Curfman, deputy editor of the *Journal of the American Medical Association* (*JAMA*), wrote, "Based on the unimpressive results to date, it will be even more important to be parsimonious about investments in stem cell therapy for heart failure unless new ideas and novel approaches open fresh avenues for potential success."

Heart Assists and Substitutes

If stem cell treatments are still over the horizon, how about mechanical assists—auxiliary booster pumps? Yes, they work, and a number are FDA-approved. Some are "bridge" devices used to support the failing heart either until it can recover adequate function or until a donor heart for transplantation becomes available. Other booster pumps are termed "destination" devices, meaning that they aim to provide long-term assistance without any anticipated additional treatment.

Three of the four general types of assist devices are for short-term use while the patient is hospitalized. The first one is essentially a heart-lung machine that sits next to the patient's bed, takes deoxygenated blood from a major vein, oxygenates it, and pumps it back into a major artery, thereby substituting for both heart and lungs. The other two just move the blood along and take some pressure, both literally and figuratively, off the heart; they are highly sophisticated catheters that the doctor inserts through an artery in the groin. One is a spinning "auger" (a.k.a. Archimedes screw) that speeds blood from the left ventricle into the aorta. The other is a long balloon that resides in the aorta and pulsates, squeezing blood ahead of it. All three of the described devices are bridges.

The fourth type could be either a bridge or a destination. It is an implantable pump that works as an auxiliary ventricle, and some of these have been in place for over ten years. Early versions caused considerable injury to red blood cells by pinching them between moving and stationary parts of the pump. Engineers have solved this problem by magnetically levitating the rotor so that it is free of any mechanical bearings and remains centered and is therefore less likely to crush blood cells regardless of the patient's position. The pump also produces *pulsed* perfusion, mimicking the natural pulse and thereby discouraging clots from forming in the pump or in the arteries. Nonetheless, it is burdensome. Patients strap the controller and power supply to their waist, change the batteries several times daily, and take great care to avoid infections around the controller-to-pump wires that pass through the skin. Biomedical engineers, bent on eliminating the driveline problems, are considering an assortment of workarounds, including small internal nuclear reactors, means of transferring power through intact skin, and using skeletal muscle contractions as energy converters.

The holy grail, of course, is a fully implantable, durable artificial heart. Investigation in this direction began in 1937, when a Soviet scientist, Vladimir Demikhov, implanted such a mechanical device in a dog, which survived for two hours after surgery. Demikhov also pioneered heart and heart-lung transplants in animals and is best remembered by his dog head transplants that resulted in two-headed bowsers.

Perhaps in an effort not to be outdone, Paul Winchell made his mark. The well-known American ventriloquist, comedian, and actor was also an inventor with some medical background. Working in conjunction with thoracic surgeon Henry Heimlich (well remembered himself for his obstructed-airway-clearance maneuver), Winchell developed a prototype for an artificial heart and in 1952 obtained the first patent for such a device, which was apparently never tested.

Robert Jarvik, a medical engineer and nonpracticing MD, was not far behind, devising iterations of similar devices. In 1982, surgeons implanted Jarvik's seventh version into Seattle dentist Barney Clark, who was suffering from severe heart failure. Clark lived for nearly four months, tethered to the 400-pound air compressor that powered the pump.

The quest continues. Now the external compressors weigh only 13 pounds, and the designs incorporate far fewer moving parts and no valves, which have reduced mechanical failures and blood clotting problems. One model is totally implantable and charges its battery through the intact skin. It is, however, too large to fit inside one-third of people who could benefit from such a replacement. Innovations on the horizon include hearts 3D-printed from various synthetic polymers. One dilemma that remains involves size versus durability. So far, the small ones are unacceptably fragile.

But getting back to Demikhov, perhaps the Soviet scientist's legacy was the inspiration he provided to South African cardiac surgeon Christiaan Barnard, who visited him twice in Moscow. After studying Demikhov's experimental results, Barnard was confident that a human heart transplant was feasible, and he performed the first one in 1967. Within a month, Norman Shumway at Stanford performed the second one, and altogether in 1968 over 100 patients worldwide received new hearts.

Immune rejection, however, plagued the early recipients, and the only available foil was high-dose steroids used for their potent anti-inflammatory properties, but they had serious side effects. Improved antirejection drugs came along in the 1980s, making heart transplants relatively safe and extending life for those with severe heart failure far longer than medical treatment alone could provide. Presently, approximately 85 percent of recipients survive at least a year, 73 percent five years, and 56 percent twenty years.

The major concern at present is that there are not enough donor hearts available for the long list of patients in need of new ones. On a faintly bright note, the one-year mortality of those on the waiting list decreased from 66 percent in the late 1980s to 32 percent 30 years later. Much of that improvement resulted from better-designed and more frequently used assistive devices.

What about getting donor hearts from another animal species, particularly one genetically engineered to reduce the risk of immune rejection? Pigs have come under the closest scrutiny, and genetic modification is underway to minimize the immune barriers. Pigs have short pregnancies and large litters and are easy to breed and maintain in a clean environment. Their hearts are appropriately sized, and the risk of cross-species disease transmission is diminished, compared to chimps or baboons, because of wider divergence of ancestry.

Yet ethical issues loom. Supporters of cross-species transplants point out that the potential societal benefits outweigh the risks, making this form of organ transfer the moral choice. Animal rights groups oppose killing animals for human use, but they seem to be less opposed to using pigs than primates.

Other caveats regarding cross-species organ transplants include risks of infecting recipients with viruses previously unknown in humans (think SARS-CoV-2 virus) and the concern that pig tissues age more quickly because a pig's natural life span is about 20 years.

Nonetheless, the scientific path forward is well defined, and cross-species heart transplants are likely on the horizon.

In face of the maladies to which our awesomely durable hearts are subject, I side with Erasmus (1466–1536), whom contemporaries lauded as the "Prince of Humanists" and who is credited with the aphorism "Prevention is better than cure." Weight control, healthy diet, avoidance of nicotine, and regular exercise are steps toward maintaining a healthy heart. We'll explore the power of exercise, contracting our muscles frequently and repeatedly, in the following chapter.

Chapter 6

CONDITIONING

REMEMBER MRS GREN, THE MNEMONIC THAT ATTEMPTS TO define life, with each letter standing for a vital function. Movement, of course, stems from muscles. Internally, those could be the smooth or cardiac types. To power ourselves around, we need the skeletal type. The R stands for reproductive functions, the domain of the genital system. Sensitivity comes from nerve endings in the skin along with the specialized receptors for sight, taste, smell, and sound, which are all parts of the nervous system. Growth hinges on the endocrine system working properly; and respiration, excretion, and nutrition come about from activities in the cardiorespiratory, urinary, and digestive systems, respectively.

Humans have easy ways to mess up every one of these functions and systems and thereby dice with their lives. Bad lifestyle choices promote disuse osteoporosis, muscle wasting, venereal disease, skin cancer, lung cancer, and alcoholic hepatitis, to name a few. Even when we try to take good care of our various systems, the direct feedback they offer us may not be anything more than a sense that something is amiss. This, of course, can be deceptive and is only confirmed or refuted by performance of a careful examination, assays of blood or urine, or by obtaining an ECG, EEG, biopsy, or imaging study. Muscles, though, communicate with us rather directly when they are not working properly.

Although we don't see our skeletal muscles as we do our skin, their contours, bulk, and tone are obvious. Furthermore, our muscles quickly reveal their functionality when we carry a bag of groceries up the stairs. And if we don't like what we see or feel or how much we are able to lift, we can condition the muscles and note progress merely by inspecting them in the mirror or by the ease with which they handle a box of books.

Another feature puts skeletal muscle entirely in a class of its own. We might abuse it massively by hiking rapidly downhill for two hours, moving furniture over the weekend, or playing 72 holes of golf in two days. For the next few days, the offended muscles are so sore that we can hardly move. Then, a week later, they are back to normal, maybe even a bit improved. Abuse your other organ systems to similar extremes and they may never recover. Also, regular exercise, within limits, stresses the skeleton and heart and stimulates them to be their best. Exercise also improves sleep, which aids health of all the organ systems.

What Happens

When you make a commitment to improving fitness, several questions naturally arise. "If I exercise more, what's going to happen?" "How long is it going to take to see a change?" "Should I lift weights or jog?" "Are there risks?" Hidden inside these questions are more fundamental ones about how muscles respond on the molecular and cellular levels when their owner sets off to improve their condition. Having a grasp of the best science available informs an exercise regimen that can be performed safely and that provides realistic expectations of improvement over time.

Think back to the image of striated muscle in chapter 5. Like very long shipping containers stacked high and wide at a seaport, striated muscle's actin/myosin units are arranged in an orderly manner, not only side to side and top to bottom but also end to end. Unlike shipping containers, however, the actin/myosin units within each cell connect at their ends. Millions of cells line up to form muscle fibers, which shorten when stimulated. To review what I described in chapter 2: the energy to power these contractions results from a chemical chain reaction. It is initiated by an electrical impulse from the nerve and ultimately knocks a phosphate off an ATP (adenosine triphosphate) molecule, thus converting it to ADP (adenosine diphosphate). Using the released energy, myosin alters its shape and advances one step along an actin filament. To repeat this process, the ADP has to be reenergized into ATP. This conversion occurs in specialized "powerhouses" within each cell. Each one is known as a mitochondrion—collectively, mitochondria (from Greek: thread + granule; first seen in insects' voluntary muscles in 1880). Mitochondria respond differently to conditioning according to the force and duration of the muscle's contractions.

Exercising against low loads over long time periods is the essence of aerobic, or endurance, activities, such as jogging and cross-country skiing. Several changes occur in response to the skeletal muscles' increased demand for nutrients during exercise sessions that are long, frequent, or both. Capillary density in the muscle increases, which facilitates oxygen delivery and thereby delays the onset of muscle fatigue. Adaptations in the cardiorespiratory system further aid exercise capacity and performance. Mitochondria in skeletal muscle cells respond by repeatedly splitting in half and increasing in number. This increases the cells' capacity to store energy-rich glycogen and fat. In the process, neither the cells nor the mitochondria change appreciably in size. Overall, endurance training builds more durable, but not larger, muscles. Aging and age-related muscle weakness are associated with a drop in skeletal muscle's mitochondrial numbers and are mitigated in part by the stimulus of regular aerobic exercise. Presently, it is unclear whether mitochondrial numbers are a controller of aging or just a marker. Either way, aerobic exercise is a foil.

In contrast to exercising against low loads, contractions against high loads over short periods characterize power, or strength, activities, such as lifting weights. In response to these endeavors, additional actin/myosin units form within the exercised muscle cells.

This stimulus causes the cells to increase in size but not in number. The mitochondria also increase their metabolic activity, which allows them to quickly convert larger quantities of ADP to ATP. Some form of progressive overload (stress) is required to effect these changes. It can be achieved by either increasing the resistance or increasing the number of times ("reps") in a training session that the resistance is encountered. In other words, the muscle responds to "time under tension" or "training volume."

The widely accepted goal is that the final repetition for exercising any muscle group should be just barely achievable, such that if one more rep was attempted, it would result in failure—dropping the weight or falling down. Despite the collective experience of exercise enthusiasts and the multitude of studies performed by muscle physiologists, exactly how this strategy leads to muscle hypertrophy is not fully understood. Batted around is the idea that the final, marginally achieved rep causes the muscle to tear and it responds by enlarging and gaining strength better to meet that demand when next encountered. Maybe "tearing" is easy to conceptualize, especially because that last rep is agonizingly hard and painful and is often accompanied by the mantra "You have to feel the burn."

There is one microscopic image from the older literature that does show post-exercise disorganization of the actin/myosin units, but current parlance is more likely to use the term "microtrauma" or "stress" rather than "tearing." I have not yet been able to discern, however, exactly what in the cell or in the actin/myosin unit proponents of the newer terminology think is being traumatized. The actual mechanism is likely a stressful combination of mechanical tension and metabolic fatigue. For the tension part, eccentric contractions, where the muscle is getting longer as it resists a weight, seem to be particularly beneficial for muscle building.

Why the mitochondria enlarge with resistance training and multiply with endurance training is unclear, and identifying any of the mechanisms responsible for either type of conditioning is hard for several reasons. First, the secrets within the muscle are shrouded at cellular and molecular levels, and human test subjects are generally reluctant to have their beloved muscles repeatedly biopsied.

Also, some of the answers may not be in the muscles but rather in the nerves leading to them, the spinal cord, or the brain. For example, multiple studies have found that exercising a muscle in one limb results in improved strength not only in that muscle but in the same muscle on the opposite side. Other studies have shown that the same cross-body strengthening effect occurs not only when flexing a muscle on one side but also by stretching it. Go figure. Although the neural mechanisms remain elusive, rehab protocols are beginning to take advantage of them. For instance, if your elbow is in a cast and the biceps is unavailable to perform resistance exercises, you might be told to exercise the opposite biceps with the expectation that some of the benefit will transfer across.

The second reason that good answers are elusive is that suitable animal models are hard to come by. Sure, scientists can aerobically train mice on a running wheel or get them to swim, but they have trouble getting those critters to pump iron and to tell the lab workers when they begin to feel the burn. Human studies are difficult to perform because of the effort of recruiting and retaining enough subjects to make the results statistically valid. The studies that are performed often use readily available college students and usually do not extend beyond three months. Would these results be reproducible in menopausal women or in octogenarians? Who knows?

The third reason is related to what I have previously described about anatomical variations from person to person. Variations among individuals also exist on a more granular level, not only anatomically but also physiologically. People's metabolisms are different, often in ways that are poorly understood. Well-recognized variables include age, sex, body size and shape, and diet. Other confounding variables related to muscle research are harder to identify and include the subjects' natural differences in their proportions of fast-twitch and slow-twitch muscle fibers and their hormone levels.

An individual's level of pain tolerance would also be a factor if the study asks them to fatigue their muscles to exhaustion or stretch a muscle beyond the point of comfort.

To minimize the influence of known and unknown variables, some studies follow the same individuals, with everyone serving as their own control. Such an experimental protocol might involve testing one method of resistance training (high reps, low resistance) for some weeks, and then, when the muscle has reverted to baseline after months of no training, test another method (low reps, high resistance).

Muscle Memory

This crossover regimen sounds good at first glance but may be confounded by "muscle memory," which has several meanings. In this context, it refers to some observations in both experimental animals and in humans that a deconditioned but previously trained muscle can be restored to glory faster the second time it is trained. It is well known that muscle cells contain hundreds to thousands of nuclei. For example, a 4-inch-long muscle cell from the human biceps contains about 3,000 nuclei whereas most cells throughout the body have a single nucleus, and red blood cells have none. Some studies have shown that, in response to resistance exercises, the muscle fibers add nuclei to support fiber enlargement. If the muscle becomes "detrained" and returns to its pre-exercise state, the added nuclei remain. Then, with retraining, the increased number of nuclei allows the muscle to (re)grow quickly and with less effort. In other words, the muscle, via the added nuclei, "remembers" its previously trained state.

Since cells' nuclei are the repository for its genes, it is from there that chemical messengers emanate to regulate cellular activity. The molecular mechanism responsible for the muscle rebounding quickly after a period of indolence may relate to how certain genes switch off and on in response to previous training. More nuclei could mean more switches are available to flip to the "on" position. Other studies, however, have not seen preservation of nuclei in formerly trained, now deconditioned muscles. This is a fertile area of research, but the exact explanation for this form of "muscle memory" is presently elusive.

Regardless of whether additional nuclei persist in trained muscle, a practical question is: how long can somebody take a break from regularly performing resistance exercises without losing muscle mass and strength? Then, if they resume weight training, can they restore their muscles to their previously trained condition? For the first question, measurable loss of strength does not occur during the first three weeks of no training. After that, muscles begin reverting toward their preconditioned state, but even in elderly people the muscles retain some of the gains achieved from weight training for as long as six months. Also, if training resumes, muscle performance can recover. Rather than the admonition "Use it or lose it," the mantra for strength should be "Use it or slowly lose most of it, but don't despair; it is recoverable."

The expression "muscle memory" has another meaning that deserves noting. People throw the term around when describing basketball players who consistently sink free throws or bowlers who can repeatedly roll strikes. Do muscles have memory to facilitate such feats? No. Although muscles are certainly responsive to conditioning, they do not get smart. Practice, however, opens neural pathways that allow the brain to send messages to muscles in a rapid, consistent, nuanced, and coordinated manner. (This is why beginning weight lifters become quickly encouraged when they see an improvement in their abilities within just the first several weeks of training. In fact, their muscle cells have not had time to hypertrophy; rather, neural pathways are responding with faster and more direct routing of electrical signals that allow more cells in a muscle to contract simultaneously.)

The so-called muscle-memory phenomenon is by no means restricted to what we normally consider athletic endeavors. Infants develop their speech "accent" by mimicking and repeating the vocalizations from the adults within hearing range. Once these neural pathways are fully established, an "accent-free" second language is often impossible to develop. Toddlers incorporate "muscle memory" to walk more efficiently. Children develop their handwriting by the same means. Mastering a musical instrument requires thousands and thousands of iterations. We may call it muscle memory, but it is actually neural pathway memory. By itself, practice will not yield perfection, but it certainly is a key element.

Peak Performance

I noted earlier that our muscle mass peaks before we are 30 years old and then slowly declines, more quickly with each decade. This is largely under hormonal control, particularly from growth hormone and testosterone. (Yes, women have some testosterone too.) Growth hormone stimulates cell reproduction and is therefore important in all aspects of human development, including muscle growth. The effect of testosterone on muscle fibers is to increase their size, which increases their strength. So it has been, is, and will be tempting to artificially elevate levels of these naturally occurring muscle builders. The side effects of doing so, however, can be life-threatening. What follows should be encouraging, because although athletic success for some activities is clearly age related, for others, it is merely a matter of picking your activity and then getting fit and staying that way.

The average age for women gymnasts winning the all-around Olympic gold medal since 1976 is less than 17. The average age for world-record holders in all sports is 26.1, and for Tour de France winners it's 28.5. Cyclists who specialize in sprinting are highly endowed with fast-twitch muscle fibers, and these speedsters peak at an earlier age. Marathon runners typically top out between ages 30 and 37. Within this age range for long-distance runners, the ascending line of experience over time eventually crosses the descending line of strength and endurance.

Clearly age matters, but it is not as restrictive as you might think. Tom Brady finally quit quarterbacking at age 45, Nolan Ryan pitched until he was 46, and tennis player Martina Navratilova retired at 50. Phil Mickelson remained a competitive golfer at 51. At the Tokyo 2020 Olympics, seven competitors were 58 or older. Equestrienne Mary Hanna, from Australia, was the oldest at 66. Among the oldsters were other equestrians, a sailor, a trap shooter, and a table tennis player. They all competed in sports that rely on slow-twitch muscle fibers and experience—tactical sports; and that is expected, because fast-twitch muscle fibers are well into decline by middle age.

Some even older athletes have achieved great heights. American Arthur Muir started climbing at 68 and summited Mount Everest at 75. Yūichirō Miura summited Everest at 70, then had two operations for irregular heartbeats before reaching the apex again at 80. I guess age was finally catching up with him because that time he needed some assistance on the descent. Oscar Swahn, a Swede, won five shooting medals over three Olympics (1908, 1912, 1920) and still holds the record for the oldest Olympic medalist in any sport at age 72. American Eliza Pollock, at age 64, earned one gold and two

A stylishly dressed athlete, Eliza Pollock, age 64, won three medals at the 1904 Olympics.

bronze medals for archery at the 1904 Olympics. By today's standards, the ever-youthful baby boomers would likely consider the ages of these athletes to be merely middling.

What about competitors who are *really* old? The shotput record for women over 100 is 13+ feet, and for men over 105 it is an even 14 feet. Remember that this event requires an explosive burst of energy powered by fast-twitch muscle fibers. (The world records for both men and women are over 70 feet, although the men's shot is 16 pounds and the women's is just under 9 pounds.) What about running events, which call on the repetitive contractions of slow-twitch fibers? For men over 100, Brit Fauja Singh, of Punjabi Indian descent, began running at 81 and holds the records for all distances from 100 meters to the marathon (26+ miles). American Gladys Burrill is the oldest woman to finish a marathon. She ran her first one at 82 and set the record for the oldest female marathoner when she was 92. Burrill became known as the Glady-ator and inspired other seniors to pick up the pace.

We also should recognize and admire the people whose athletic successes do not include world records, Olympic medals, or even finish lines. Dancers, for instance, exhibit amazing athleticism. Vocalists, musical instrumentalists, and even sign language interpreters may not look fit in gym shorts and tank tops, but their performance muscles must be as well trained and as carefully nurtured as those of any track star.

Despite all the good that comes from exercise, our "civilized" lifestyle works against us. The Industrial Revolution was a major contributor to humankind's muscular decline because machines replaced much physical labor. Also, people moved off the farms and into the cities. Today hunting and gathering consists of driving to and from the grocery store or making restaurant reservations.

In multistory buildings, the location of the elevator lobby is obvious, while the location of the stairway is not. Laborsaving devices are ubiquitous. Not only do we drive rather than walk or cycle, muscle-sparing features of today's cars themselves are seductive and insidious. Automobile-induced indolence includes automatic garage door openers along with electric starters and power brakes, steering, windows, and seats. At home, Alexa, Roomba, and the TV remote mean more time on the couch. Have you milked any cows lately or churned any butter? I could go on, but you get the picture. It should be no wonder that in the past 40 years, the percentage of overweight Americans has increased from 50 to 70 percent.

An improved diet would help, but it is a cruel trick of nature that the pleasure areas in the brain are maximally stimulated by my three favorite food groups—butter, salt, and sugar—a fact not lost on manufacturers of convenience foods.

Age Matters

By the time we are 70, a fourth of our muscle mass has wasted away. For those lucky enough to make it to 90, only 50 percent of their muscle mass remains. That is more than enough to increase risk of death from pneumonia that is secondary to weakness in breathing and coughing. Even a 30 percent loss of muscle mass equates with the loss of the ability to live independently. Balance and agility may dissolve faster than strength, because fast-twitch muscle fibers,

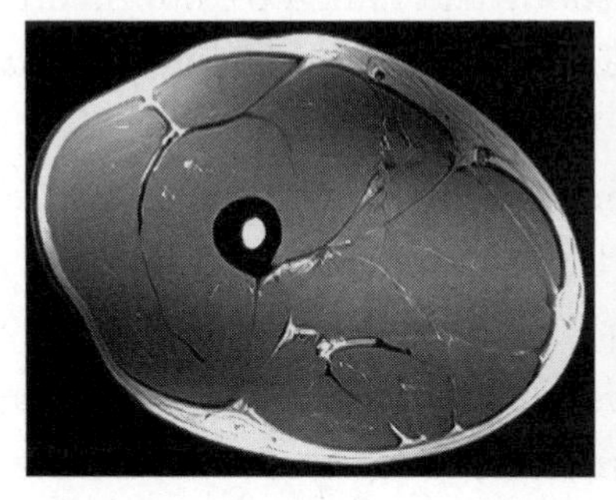
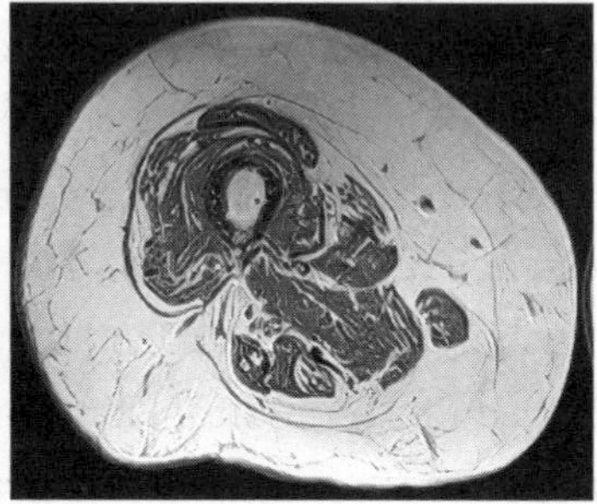

These are magnetic resonance images taken through the mid thigh. In both frames, the central black ring is the thigh bone, and the peripheral white areas are fat. The intermediate shade is muscle. Left: A healthy 33-year-old with well-developed musculature. Right: A 66-year-old with marked muscle atrophy.

responsible for quick reactions, diminish in size and number more rapidly than slow-twitch fibers. And just a 20 percent loss of overall muscle mass increases the risk of falling, which, if the hip breaks, has fatal consequences within a year for 20 percent of victims.

Is there hope? Absolutely, but we are going to have to resume some habits of our hunter-gatherer ancestors. Worse, we may have to pay money for exercise opportunities that they got for free just in the process of surviving. Payment may range from the price of walking shoes and some exercise bands to gym membership or a personal trainer. Is it ever too late? No. Multiple studies looking at seniors at home, in nursing homes, and even in hospitals found that all responded favorably to resistance training.

For example, one study followed 72 people at least 65 years old who had mild to moderate risks of falling. In a randomized controlled manner, they underwent training for body awareness, strength, reaction time, and balance and showed significant improvement in all categories. Additionally, and perhaps not surprisingly, they also experienced improvement in overall health-related quality of life.

Another recently reported study compared testosterone levels and body leanness between endurance-trained middle-aged athletes and age-matched untrained individuals. We shouldn't be surprised that the athletic group won on both counts. In other words, increasing muscle activity and strength promote and perpetuate the MRS GREN life functions in many ways.

Finding Fitness

Today, fitness programs beckon. Studio and gym sessions as well as online options include Jazzercize, step aerobics, water aerobics, Zumba, Pilates, yoga, Curves, Spinning, rock climbing, parkour, and tai chi. Consider too that the five-times-daily Muslim prayers, which include repeatedly standing, kneeling, and bowing, provide general toning and stretching of major muscle groups.

If you want an adversary to spur you into fitness, consider Tae Kwon Do, judo, karate, Krav Maga, boxing, kickboxing, or chess boxing. The last adds an intellectual element by alternating six rounds of chess with five of boxing, each three minutes long. When time is even more limited, or you really want to keep your heart pounding to get a cardio workout along with your resistance training, consider high-intensity interval training (HIIT), where you exercise one major muscle group strenuously for 30–40 seconds. Then, following 10 seconds of breath catching, you stress another group while the recently stressed group rests, albeit briefly. Performing several "circuits" of 7 to 10 prescribed exercises will leave you sweating. Smartphone apps, several of which require no weights (other than your body weight), make it possible to have an HIIT session anywhere you go. And don't overlook unsupervised simple pleasures such as jogging, rapid walking, dancing, biking, and swimming.

Regardless of your chosen activity, here are three pieces of advice.

The first one may save your life. If you are not in the habit of exercising regularly, clear the plans with your doctor. To paraphrase what the commander told Tom Cruise in *Top Gun*, "Don't let your ego write checks that your body can't cash."

The second piece of advice will extend your athletic longevity. Exercising under the eye of a professional or experienced friend, at least until you fully understand your limits, will help you make steady progress and minimize risk of discouragement or injury.

The third one also relates to performance and injury. No more than you should shout at a sleeping guard dog, you shouldn't startle your muscles into maximum performance without first warming up. This is intuitive for a lot of typical athletic activities, because as spectators at sporting events, we see the participants ahead of time preparing for all-out performances by progressively awakening their neural networks, muscles, and joints and getting more blood flowing to all these areas. So when we exercise, we follow their lead, or at least we should. Warm-up should also precede resistance training and even be an antecedent to moving the fridge or otherwise contracting our muscles forcefully. Singers and broadcasters also warm up, but they usually do so in private. Therefore, it is harder for others to know and emulate their warm-up routine to avoid mincing words and straining notes by suddenly calling on sleepy vocal muscles. Exercises, taught to me by a singer, that I perform before making an oral presentation include some forceful "huh, huh, huh"s to awaken the diaphragm, repeating "red leather yellow leather" for a minute or so to really get my throat and tongue moving, and making buzzing sounds with my lips while humming.

Regarding resistance training, how often and how much is right once warm-up is complete? The American College of Sports Medicine advises training with weights two to three times weekly on non-consecutive days, although studies have shown that lifting weights even once a week has salutary effects. Typically, a "set" consists of eight to twelve reps for exercising one muscle group, biceps curls, for instance, followed by two to three minutes of rest, followed by several more sets. The goal is to be only marginally able to execute the last rep of the final set. Struggling to complete the last concentric contraction seems to be particularly helpful in getting the muscle's attention. The "burn" one feels is the muscle telling itself that it needs to shape up before the next training session. Even the wizards do not fully understand what is going on metabolically and microscopically, but the stressed muscles do respond to the amount of time under tension, which has to do with the number of reps and sets and the load resisted.

Regimens for resistance training can be complicated and potentially injurious, so find somebody experienced to help figure out what's a good plan for you. Tell the trainer at the outset what your goal is: size, strength, or endurance.

In addition to regular resistance training, the American College of Sports Medicine and the World Health Organization advise regular aerobic exercise to get your heart rate up for at least 150 minutes a week. This directly benefits your cardiovascular system in at least three ways. It improves the heart's muscle tone, lowers its resting rate, and decreases the tendency for artery-clogging atherosclerotic plaques to form. Such exercise could come from a moderately brisk walk three to five days a week or a high-intensity run less often or for less time.

Aerobic activity also improves bone health by indirect means. This occurs because feet impacting the ground send jarring messages to the bones in the lower extremities, pelvis, and spine and stimulate them to shape up. What happens is interesting. The calcium crystals of bone respond to mechanical stresses by generating minute electrical charges. Bone-forming cells sense these charges, which are coming from areas where the bone is being stressed and in need of at least maintenance if not reinforcement. The bone-forming cells go to work and strengthen those areas, which for walking and jogging are the spine, pelvis, and lower extremities. But walking and jogging stress these bones in a predictable, one-directional manner and do not stimulate adaptive skeletal responses in the upper limbs at all. Resistance training, however, provides diversified loading throughout the skeleton and stimulates generalized bone health. As good as cycling and swimming are for cardiorespiratory fitness, they do not foil bone wasting (osteoporosis) in the way that brisk walking and jogging, and especially resistance training, do.

Do you need to walk the widely discussed 10,000 steps a day to maintain strong bones and a healthy heart, both of which are integral to the optimal performance of skeletal muscle? Here is how this not-exactly-scientific recommendation came about. Tagging onto the interest in fitness generated by the 1964 Tokyo Olympics, a Japanese clockmaker started manufacturing and marketing a pedometer, the name of which looked like a walking man when written in Japanese characters. The same characters also translated as "10,000-steps meter." (Imagine where we would be if it translated as "all-day-on-the-sofa meter.") Later, modern fitness monitors picked this nice round number as a goal, which amounts to about five miles and requires almost two hours for most people to accomplish, not necessarily an achievable goal for all. Will less suffice?

A 2021 study provides strong evidence that even less ambitious stepping is quite beneficial. The study pooled data on over 28,000 subjects from seven recent investigations. The baseline group took 2,700 steps per day. Above that, the risk of mortality dropped 12 percent for each additional 1,000 steps in the general adult population and 13 percent for adults over 70. So 10,000 is great, but each additional 1,000 steps (which might take nine to ten minutes) over a baseline of 2,700 is beneficial.

For basic step counting, your smartphone is all you need. Apple iPhones come with the *Health* app installed. Tap its heart icon and then touch "Activity," and you can see your steps recorded by day, week, month, and year. Android device users may have to install Google Fit from Google Play and then launch the *Fit* app. Then, of course, keep your phone with you as you move about. Marching in place doesn't work, and cycling gives you only partial credit for distance traveled. It's fun to see the numbers and graphs and know that you are being proactive with your health, most directly for the cardiovascular system but indirectly and equally important for your musculoskeletal system.

After performing either resistance or aerobic exercises, the next advice from the American College of Sports Medicine is to spend several minutes stretching all major muscle groups while they are maximally warm and pliable and therefore well protected from injury. Muscle contains a lot of water, which leads to several comparisons. An icicle is brittle and easily snapped into pieces. Likewise, when your fingers are really cold, they are clumsy and feel as if they could snap off. The extreme case of this response to cold occurs with accidental total-body immersion in icy water. The autonomic nervous system constricts the arteries supplying the limbs to keep the vital organs working. The muscles become stiff and useless in about 10 minutes. So cold is clearly bad for muscle function. Rather than ice, think about water in its warmer, liquid form. Belly-flopping into a pool stings because the water resists being suddenly deformed. Conversely, the water offers no resistance at all when you slowly step into a pool. It is the same with the water-rich muscles as well as their stretch receptors. Jerk them, especially if they are cold, and they object and may resist the stretch to the point of pain or even tearing. Warm them up and move them slowly, and they comply. Remember, no shouting at sleeping guard dogs.

Furthermore, studies show that stretching muscles to their limits temporarily weakens them. So it is counterproductive to perform stretches pre-exercise. Before I knew these study results, I had discovered this phenomenon on my own. I used to jog about three miles each way to an early morning yoga class. All the yoga stretches felt great because my muscles were fully warmed up, but heading home afterward, my legs felt as if I were running through molasses.

You might next ask, "Why stretch at all?" Doing so draws muscles out to their maximum length and maintains full capacity of the joint capsule, which surrounds the joint (as is depicted in figure at the beginning of chapter 3). When both the muscles and the joint capsule are fully supple, the joint can move through its entire natural range of motion. At rest, muscles and joint capsules tend to gel, and then they offer some resistance when subsequently stretched. The longer they rest in one position, the more they resist being disturbed. The muscles may feel stiff or creaky when getting out of a car after a long trip or getting out of bed in the morning, particularly if the muscles and joint capsules were well exercised the day before. If muscles and joint capsules are not put through their full paces for weeks or months, their shortened positions can easily become the new normal. Then it is hard or impossible for them to regain their youthful suppleness. In other words, use it or lose it.

After warming up, exercising, and stretching, how important is a cool-down session? Can it be a passive activity such as standing around chatting, sitting in the locker room or sauna, receiving a massage, or posing as a corpse (the savasana, often the finale of a yoga session)? Or should the cool-down mildly engage the exercised muscles with a walk, a slow swim, or some light weights? What happens if muscles aren't gently cooled off and are rushed onward into the owner's busy day? Everyone seems to have opinions regarding answers to these questions, yet there is sparse scientific evidence to support most recommendations.

Sports scientists Bas van Hooren and Jonathan Peake recently surveyed the available peer-reviewed literature in 145 journal publications. To summarize their findings: An active cool-down such as walking or light jogging does little to improve athletic performance later the same day or over the next several days. This may be in part because the active cool-down interferes with efficient restoration of the muscle's glycogen energy stores. Lactic acid, a metabolic by-product of strenuous muscle activity and traditionally thought to be a contributor to fatigue, may be removed from the blood more rapidly with an active cool-down, but it does not necessarily disappear more rapidly from the recently exercised muscles. An active cool-down does not reduce the often-encountered post-exercise joint-motion limitations and muscle stiffness, nor does it reduce the incidence of subsequent injuries. Nonsteroidal anti-inflammatory medications and cold-water immersion treatment are counterproductive because they blunt muscle's ability to respond to inflammatory stress by increasing strength or endurance.

Overall, the collected science shows few physical benefits of actively cooling down, so if time is an issue, it would be far better to spend precious minutes warming up before exercising and stretching afterward. Nonetheless, an active cool-down is a chance to relax, brag, tease, and reflect on the recently accomplished athletic endeavor—psychological benefits that should be recognized.

The same level of uncertainty surrounds the purported benefits of massage and foam rolling to aid recovery after exercise. Certainly the mechanical stimulation feels good, and the psychological benefits cannot be overlooked, but studies offer conflicting or inconclusive results.

Another area where hype at least runs even with the supporting science is the use of compression garments both as aids to athletic performance and to recovery. It started in 1958 when a DuPont chemist invented a synthetic fiber capable of stretching up to five times its resting length before easily returning to its prestretched form. It is known as *spandex* (one trade name is Lycra) in the United States and as variants of *elastane* in many other countries. The French ski team introduced these elasticized garments to the sporting world at the 1968 Winter Olympics, and the fabric's meteoric rise in popularity occurred in the 1980s, when the availability of spandex garb paralleled the blossoming interest in physical fitness generally. Now, colorful socks, sleeves, shorts, tights, and full bodysuits made from spandex are popular among both recreational and competitive athletes.

How might these compression garments aid muscle conditioning? Claims include facilitating the circulatory system's job of delivering oxygen and nutrients to the muscle and aiding the removal of metabolic wastes. These effects could theoretically reduce muscle damage and hasten recovery. An elemental problem, however, is that nobody knows what the right pressure should be. Logically, it should be higher near the ankle and lower further up the leg. Even if the proper pressures were known, achieving them with anything less than custom-made garments is unlikely.

Nonetheless, numerous studies have tried to tease out compression garments' effect on both endurance and resistance activities. Because the researchers used individuals of different ages and studied them for different lengths of time during and after exercise while measuring different factors, the results are, not surprisingly, mixed and confusing. The consensus, however, is that compression garments do little if anything to enhance performance, but neither do they detract. Whatever improvement an athlete senses may simply be a placebo effect. Of course, blinded studies including a control group are impossible because in an instant subjects would know whether the experimental socks were squeezing their calves.

If compression garments do not clearly enhance performance, what is their effect on recovery? An answer to this question would be particularly relevant to an athlete who is in a tournament and competing several times in a day and possibly over several consecutive days. Here results pooled from multiple studies provide better evidence, particularly for resistance activities. Strength recovery, more so than endurance recovery, seems slightly better with the use of compression garments. Whether this is better than massage or foam rolling is untested. My bottom line on compression garments is that they look cool, do not detract from performance, and may aid recovery a bit, but more time spent warming up, exercising, and stretching is likely better than squeezing oneself into tight clothes.

In addition to exercising and stretching regularly, the American College of Sports Medicine's final advice regarding conditioning is to regularly work on balance, agility, and coordination. These skills, which are inherent in exercise programs such as tai chi, Jazzercize, yoga, and Zumba, may not make us graceful, but if they keep us from falling over and breaking something, then they serve their purpose. Even if you don't want to go dancing, an easy way to work on these skills is standing first on one leg then the other while brushing your teeth. If that seems easy, try closing your eyes momentarily. You also can get a twofer by unobtrusively standing on one leg while waiting in line or for an elevator.

Better yet, take the stairs and work in some aerobic, resistance, stretch, balance, and agility training all at once. Multiple studies have demonstrated the benefits of doing so. Ascending one flight increases longevity by ten seconds, which means that if you can get yourself up there in less than ten seconds, you have a net gain in life span. You may ask how they figured this out. One study measured the longevity of London double-deck-bus drivers (sedentary) versus conductors (repeated stair climbers). Another study looked at the longevity of ground-floor apartment dwellers versus those who walked up eight flights to get home. A third investigation followed 4,000 Harvard alumni for 20 years. Those who climbed at least 36 flights of stairs a week had an 18 percent reduction in mortality compared to those who climbed less than 10 flights. Even for people with known or suspected cardiovascular disease, those who could maintain or recover the ability to climb four flights of steps in a minute cut their risk of death from all causes over five years by half.

Training Aids

For individuals rehabilitating from an injury, stroke, or surgery and for whom a regular workout or even stair climbing would be too painful or stressful, there is another way to stimulate muscle development without forcefully exercising. It is almost magical. At least it's humbling, because nobody is sure how blood flow restriction (BFR) works. It starts with placement of an inflatable blood-pressure-type cuff around the thigh or upper arm and inflating it to a pressure sufficient to restrict arterial blood flow into the limb and completely stop the low-pressure venous blood flow returning from the limb to the heart. Then the participant performs resistance exercises with that limb—knee straightenings or biceps curls, for instance. The effort is low-intensity because the resistance is only 20–40 percent of the weight that person could normally handle for one rep under ordinary conditions. With BFR the person does 10 to 15 reps in each of three to four closely spaced sets. The last rep in the last set should be as hard to achieve and should create as much "burn" as would normally occur with far heavier resistance and no cuff.

Doing this two or three times weekly enlarges and strengthens not only the exercised muscles but also, surprisingly, the muscles in the opposite limb and in the trunk. In other words, by retarding blood flow, accumulated metabolic by-products somehow stimulate muscles near and far to respond as if they had all experienced high-intensity workouts. BFR works even for a person who cannot lift weights at all—for example, someone whose muscles are in a cast or who is bedridden.

The technique was invented in the 1960s in Japan. It remains under active investigation regarding the fundamental mechanism along with the optimal dosage parameters including cuff width and pressure. Since the participant needs to exhaust the exercised muscle and experience discomfort while struggling with the last rep, BFR is no shortcut for an entirely healthy individual to get buff. The light resistance used in BFR, however, puts less strain on the heart, bones, tendons, and joints, so individuals with limited capacity can nonetheless recover muscle function. Intentionally slowing blood flow, however, risks blood clots, so BFR is typically ill-advised for people with a history of clots and for seniors.

Another "shortcut" that has been around longer than BFR is electrical muscle stimulation (EMS), which had its origins in Luigi and Lucia Galvani's lab in 1780, where they stimulated frog legs to twitch in response to an electric shock. In the 1960s, Russian sports scientists applied EMS to elite athletes and claimed 40 percent gains in strength. As with many aspects of muscle-performance investigation, the hype ran ahead of the science.

In the United States a classic example was the commercially available Relax-A-Cizor. The user applied electrodes to the skin surface over a set of muscles and flipped the machine's switch, sending electrical currents sufficient to vibrate the muscle 40 times a minute. One advertisement stated, "If you want smaller hips, take a Relax-A-Cizor to bed with you." Another touted, "New Relax-A-Cizor invents sexy exercise for the man." The device's reach turned out to be farther than its grasp, because by 1970 the Food and Drug Administration had heard enough about associated blood vessel ruptures, hernias, cramps, nervousness, miscarriages, loss of consciousness, and a panoply of other symptoms. The FDA not only banned it but also prohibited secondhand sales and recommended that owners destroy or otherwise disable the machines.

Fast-forward to the present time. EMS is alive and well. On Amazon you can search for devices by price category from those under $25 to ones over $200. Those are the over-the-counter ones. Others are by prescription only, make more modest claims, and have legitimate applications. At the FDA's insistence, none claim weight reduction. They are useful in rehabilitation situations when musculoskeletal injuries curtail conventional weight training. As with blood-flow restriction, EMS may also be helpful for bedridden individuals.

In 2021, the FDA approved an EMS device to combat snoring and mild sleep apnea, conditions in which the relaxed tongue with its eight muscles collapses rearward into the airway, causing obstruction. A U-shaped device, akin to a lower denture, fits on either side of the tongue, with the controller and battery hanging out and down over the chin. In 20 minutes a day, it tones the tongue so that it behaves when its owner is asleep.

Also available are several products to help women overcome urinary leakage when sneezing, laughing, or lifting heavy objects. These are pads or bicycle shorts that have embedded electrodes that stimulate the pelvic floor muscles to contract and strengthen in the same way that performing a Kegel contraction is intended to do.

Rather than trying to train muscle by stimulating it electrically, how about activating it mechanically, by vibrating it? The idea of standing on a vibrating platform to enhance bone and muscle strength has been around for a long time and theoretically has appeal, particularly for astronauts in prolonged low-gravity environments and for elderly individuals who might not be otherwise able to perform resistance exercises. For the rest of us, numerous studies have been published regarding whole-body vibration therapy, but the results are mixed and are by no means conclusive regarding any benefits not available through diet control and conventional exercise. That has not slowed the commercial promotion of such devices. Advocates claim that 15-minute sessions three times weekly aid weight loss, burn fat, reduce post-exercise muscle soreness, and build strength. Let the buyer beware.

Nutrition

Exercise alone will not result in well-conditioned muscles. Of course, diet plays an important role. In particular, proteins, which are long chains of amino acids—the building blocks of life—count. Actin and myosin are proteins, as are many of the molecules that mediate nerve conduction, neuromuscular junction activity, and intramuscular signaling that lead finally to a muscle contraction. To build muscle, adequate protein intake is paramount. Recall that resistance training results in muscle cells adding actin/myosin units, which makes the cells larger and the muscles bulge. The building materials for those additional actin/myosin units are amino acids, which come from ingested proteins. Three unfortunate realities cloud the issue.

First, salt, sugar, and fats provide a buzz just not available from a helping of protein-rich jerky or cottage cheese.

Second, you probably saw Sylvester Stallone as "Rocky" start his day by cracking several raw eggs into a glass and gulping them down before starting on his predawn runs. Again, that's not a particularly appealing breakfast for most of us, and a lot of solid science has come out since Rocky was first romancing Adrian in 1976. At the time, Rocky was basing his diet on "bro science" gleaned from his jockish "bros" at the gym. Without any basis beyond their intuition, which turned out to be faulty, they thought, "If some is good, more must be better."

The third issue is one of measurement conversions. Research papers typically report recommended protein intake in grams of protein per kilogram of body weight per day. That makes sense for scientific communications but probably causes your eyes to glaze over. For metric-system-dim Americans, who have little awareness of what a gram is or their weight in kilograms, the recommendations are gibberish (except for the *day* part, which we understand). Here I convert the recommendations based on body weight from kilograms into pounds. For someone who is physically inactive, 3.6 grams of daily protein intake per 10 pounds of body weight is sufficient. For generally active people who are not particularly interested in building muscle, 6.4 grams per 10 pounds of body weight is good. If you want to build and maintain muscle, then the needed construction material in the form of ingested amino acids should be increased to 7.3 grams for every 10 pounds of body weight, which is just over twice the recommended amount for couch potatoes.

The next part of dietary planning requires knowing how many grams of protein are in a helping of various foods. Tables listing this information are easily accessible. In short, meat, fish, milk, yogurt, cottage cheese, beans, and nuts are protein-dense.

As an example, a person weighing 140 pounds who wants to be in shape for jogging needs 90 grams (140 pounds divided by 10 times 6.4) of protein daily, which could come from two helpings of meat, a glass of milk, and egg and toast, and one serving each of beans and quinoa. The same person who wants to get the most from weight training should consume 102 grams of protein (140 pounds divided by 10 times 7.3) every day, best divided among four meals. This is still easily achievable, even for vegetarians (no meat, poultry, fish, seafood) and vegans (no meat, poultry, fish, seafood, dairy, or eggs). The takeaway is that it is not hard to have a high-protein diet.

But does every athletically inclined individual need to consume all that many amino acid chains every day? Colleen McKenna and colleagues provide the most current answer in a 2021 article in the *American Journal of Physiology—Endocrinology and Metabolism.* They randomized 50 adults ranging in age from 42 to 58 into two groups. One consumed a moderate-protein diet for ten weeks, the other a high-protein one. During that time both groups performed supervised weight training three times weekly. After analyzing the data, McKenna and associates concluded, "Higher protein intake above moderate amounts does not potentiate resistance training adaptations in previously untrained middle-aged adults." In other words, middle-aged people interested in building muscle bulk and strength can do just fine on a moderate-protein diet.

So the bro-science mantra about needing a high-protein diet is hooey, right? Not so fast. There are rarely absolutes in science. First, the McKenna study lasted ten weeks. Would there have been differences if the study had lasted a year? Maybe, but such a study would be exorbitantly expensive, and if several subjects dropped out, the additional data collected over a longer time would be skewed and worthless. Next, the study subjects were, in McKenna's words, "at the onset of aging." Would the results be the same for younger or older individuals? The subjects had not previously lifted weights. Would "trained" subjects have reacted to a moderate- or high-protein diet differently?

I dwell on such unknowns in the McKenna paper because it exemplifies the lack of clear applicability that permeates the literature not only on exercise and diets but also on supplements, discussed next. The perfect study that has universal applicability has not been done, and it would be impossible to do. So some extrapolation from available data is likely reasonable, but it is best not to commit any one study to your "eternal truth" folder. Rather we should seek a consensus of all available information, which continues to grow.

The bottom line for me, since my Olympic and Mr. Universe aspirations and opportunities have passed, is that a moderate-protein diet will most likely meet my needs. Nonetheless, from what I have learned, I am now more likely to choose salmon or scallops from a menu over pasta. (That is with one exception. Should I lose my senses and ever run a fourth marathon, I would "carbo-load" for several days before the race to maximize my glycogen stores.)

Speaking of pasta, can vegetarians and vegans build muscle and be athletically fit? Certainly, and a number of elite and record-setting athletes do shun all animal protein, and others (pescatarians, for whom fish and seafood are okay) avoid all *land-based* animal protein. These restrictions do, however, risk undersupply of essential micronutrients including iron, zinc, iodine, calcium, vitamin D, and vitamin B12; so their diet and any nutrient supplementation need to be carefully structured, ideally under the supervision of a sports nutritionist.

Supplements

Even for omnivores who could potentially derive from their food all the calories they need—along with essential amino acids, fat, minerals, and vitamin nutrients—nutritional supplementation beckons seductively. In 2000, the dietary supplement industry was worth $120 billion and likely will triple in value by 2027. That is similar to the gross domestic products of Finland and Vietnam. How much of that is necessary or beneficial? Maybe the pill, powder, or elixir does not even contain the said substance. Maybe it contains a concoction that also contains unnamed substances, potentially toxic.

The uncertainty stems from the fact that the FDA does not control vitamins and other over-the-counter nutritional supplements as it does prescription medications, so supplements are the Wild West of commercial opportunity. If a company touts it, somebody will likely buy it. Since consumers have paid for it, they will be inclined to note a benefit, even if it is no more than a placebo effect. Supplements for improving every aspect of MRS GREN abound. Is there any science behind their use for enhancing muscle performance? Some, for at least two elixirs.

The benefit of adequate protein intake is undisputed; and with respect to weight training, the prevailing advice is to give the muscles a 20-gram protein boost when they are maximally hankering for it—right after a workout. Look at a table showing the protein content of various foods and find something that fits into your diet and that is also palatable, not highly caloric, and easily accessible. Hence a steak filet on the way to work is probably too messy, and most protein bars are highly caloric. Jerky works, if you like to gnaw. Protein powders that you mix with water are also convenient. Whey powder (a by-product of the manufacture of cheese) comes in various flavors including chocolate, cookies-and-cream, and strawberry. Yum. Vegan alternatives made from peas, beans, and seeds are also available.

Solid evidence and vast experience also support supplementation with creatine. This rather simple molecule, consisting of two amino acids, is naturally present in high-energy-demanding tissues, particularly muscle and brain. In muscle, creatine can quickly yield its energy to ADP, thereby converting it to ATP, which is then available to power another myosin/actin interaction. Multiple studies show that creatine supplementation increases muscle mass and strength after high-intensity exercise. It does so by allowing the participant to perform a greater volume of work (reps × sets × load) before fatigue sets in. It also hastens post-exercise recovery. Hence, creatine allows for more exercise more frequently, and the muscles respond accordingly. Seniors and untrained individuals of all ages benefit along with devoted lifetime jocks. Furthermore, creatine's safety profile is outstanding, and over decades of scrutiny in numerous studies, no adverse effects have surfaced. Additionally, creatine is entirely legal. It is present in meat, but in such small quantities that you would have to eat several pounds a day to see an effect. Far more practical is to mix the faintly bitter powder with water or include it in a drink of flavored whey powder.

Now that you know the benefits of boosting the protein and creatine in your diet, how about stirring in some fertile-egg yolks and dark chocolate? To understand the rationale for this suggestion, it is important to know a little about myostatin. It is a protein produced by skeletal muscle cells. Myosin puts the brakes on muscle growth, particularly at the end of adolescence. Before then, myostatin levels are normally low while growth hormone and, in boys, testosterone levels are high. This combination accounts for the typical growth spurt seen in teenage boys and accounts for the ease with which they can bulk up. Toward the end of adolescence, the muscles produce more myostatin, which suppresses continued rapid muscle enlargement.

Strangely, Belgian Blue cattle are missing the gene for myostatin, so their muscles continue to grow appreciably larger even in adulthood. Although these cattle are described as "double muscled," they in fact are about 40 percent larger than normal. Either naturally or because of genetic engineering, there are breeds of dogs, rabbits, goats, and pigs that also overgrow their muscles due to the lack of myostatin's inhibiting influence.

Mice with myostatin deficiencies have spent time in a low-gravity environment on the International Space Station. They return to earth

A naturally occurring genetic deficiency of myostatin has made this eight-month-old bull extremely muscular.

fully muscled, which is remarkable considering that in the same setting the skeletal muscles in astronauts and mice with normal levels of myostatin waste away. Perhaps even more remarkable is that the myostatin-deficient mice also maintain their bone mass in space, apparently because their large muscles are stressing the skeleton, which responds by adding new bone. Hence, it might be helpful for astronauts to take a myostatin inhibitor on long space trips in order to maintain muscle mass and bone strength.

Rarely, humans either lack the gene that produces myostatin or their muscles are genetically blocked from responding to myostatin's braking function. Hence a 10-year-old boy may have a child's cherubic face and a physique ready for a miniature bodybuilding contest. Of course, some adult bodybuilders want to know how they can disable myostatin's natural braking action. Two compounds known to inhibit myostatin attract their interest. In small quantities, they are present in dark chocolate, green tea, and chicken egg yolks, especially fertilized ones. There is no convincing evidence that diets supplemented with any of these myostatin inhibitors work to suppress myostatin effectively. In 2019, the World Anti-Doping Agency banned use of these compounds in their purified, injectable forms but did not clearly indicate the legality of dietary sources.

There is excellent scientific support, from both laboratory and clinical studies, for the dietary benefits of protein and creatine on muscle maintenance and development. There is also good support for the effect of myostatin inhibitors in laboratory settings. Beyond that, the scientific rationale for taking other supplements is shaky; and even if one or two well-executed studies should show a statistically valid 5 percent gain achieved by weight-trained young men over ten weeks, you should ask three questions. First, why did the study test just physically fit collegians and for just a short period? That answer is easy—the study was doable, both logistically and financially. Next, wonder if the results can be translated to women, other age groups, or untrained individuals. And finally, even if the results might translate to other groups, is a 5 percent gain worth the expense and possible risks for an avocational athlete? This is where bro science kicks in. Competitive athletes looking for the slightest edge to achieve fame may have a gym bag or kitchen cabinet stuffed with supplements and "enhancers" and swear by their effectiveness. Maybe they are just experiencing a placebo effect. Maybe it calms them to know that they are doing *everything* possible. "I know Joe is taking 'X,' I'd better take it too." By contrast, I just want to stay reasonably fit, be healthy, and have fun. I'd rather spend my money on a tasty steak.

Performance Enhancers

By what other means might a competitor gain an edge? Plenty, and here is a brief timeline. By the third century BCE, Olympians and gladiators were using brandy, wine, mushrooms, sesame seeds, and herbal tea as some of the first performance-enhancing drugs (PEDs). Doping took a far more serious turn in the late 1800s when endocrinologist Charles Brown-Séquard started injecting himself with testicular extracts from guinea pigs and dogs to restore vitality. The doctor noted that taking

Left: Major League Baseball player and Hall of Famer Pud Galvin pitched for the Pittsburgh Alleghenys (now known as the Pirates) in the late 1800s. He is the first baseball player widely known to use a performance-enhancing drug—an extract of monkey testosterone.
Right: Thomas Hicks won the marathon at the 1904 Olympic Games in St. Louis. After flagging in midrace, his supporters revived him with two strychnine injections and a drink of brandy and raw eggs.

what was to become known as the Brown-Séquard elixir increased his physical strength, mental abilities, and appetite. (The active ingredient in the elixir was unpurified testosterone, which causes enlargement of muscle cells, thereby increasing their strength.)

Pitching for the Pittsburgh Alleghenys, James Francis "Pud" Galvin openly used Brown-Séquard's potion and held the Major League Baseball record in 1889 for wins, starts, complete games, and innings pitched. A newspaper reported that Galvin was "the best proof yet furnished of the value of the discovery." His nickname came from making hitters look like pudding, yet nobody seemed to care much about his, perhaps onetime, use of a PED. One of the nineteenth century's greatest pitchers, Galvin was finally elected to the MLB Hall of Fame in 1965.

During the 1904 Olympic marathon, faltering Thomas Hicks received two strychnine injections along with oral supplements of brandy and raw eggs. (Strychnine works by markedly amplifying the conversion of a nerve's electrical impulse into the muscle's chemically mediated contraction. In small doses, strychnine gives a boost to fatigued muscles. In larger doses it boosts them into convulsions and leads to death.) The amount that Hicks received revived him sufficiently that he went on and won the race. This and similar upfront instances of doping led the International Amateur Athletic Federation to ban stimulating substances in 1928, but the ban was unenforced and ineffective.

Soon after the artificial synthesis of testosterone in 1935, problems with PEDs multiplied. (Male guinea pigs and dogs were undoubtedly relieved.) An observer in the 1950s noted, "You need never go off-course chasing the peloton [pack of cyclists] in a big race. Just follow the trail of empty syringes and dope wrappers." Ten years later, shotput, javelin, discus, and hammer throwers were taking testosterone, which promotes muscle growth with an attendant increase in muscle strength of 5–20 percent, but it does not increase endurance. It also shrinks testicles.

The first Olympics drug testing took place in 1968, at the winter games in Grenoble and the summer games in Mexico City. The screening must not have been particularly effective, because in the 1970s, it turns out, the East German athletes, who seemed to dominate the awards podium, all took testosterone-like muscle-enhancing steroids by mandate. Sailors were excepted, maybe because the sport requires more mental than physical strength for success. Canadian sprinter Ben Johnson, "the world's fastest man," also distinguished himself in another way. In 1988, he was the first to have his Olympic gold medal stripped away after testing positive for a testosterone derivative.

To combat PED chicanery, the World Anti-Doping Agency (WADA) came about in 1999 and has been fighting an uphill battle from its inception. A drug is illegal if it meets at least two of these three criteria: improves performance; poses a hazard to the athlete's health; and violates the spirit of sport, which should be a celebration of the human body, mind, and spirit. If you want an eyeful of what resourceful chemists, pharmacologists, trainers, and athletes are working for/against, search online for the *WADA Prohibited List*. Many of the forbidden drugs are synthetic forms of testosterone and growth hormone, both of which have direct effects on enhancing strength.

Some substances are allowable if they are present only in small quantities. An example is caffeine, which reduces fatigue and hastens post-exercise recovery. It is a smooth-muscle bronchodilator and therefore in high doses would improve breathing for endurance events. WADA allows inhaled vasoconstrictive agents, used for asthma, when their concentration does not exceed established thresholds and when the athlete has a therapeutic-use exemption. Are these "asthmatic" athletes getting away with something? Apparently not. A recent study looked at all Olympic competitors from 2010 through 2018. Less than 1 percent of athletes had therapeutic-use exemptions. The study's authors found no correlation between possessing such an exemption and winning a medal.

Why do athletes dope to begin with? Sports are important for an individual's physical and mental development and promote international cooperation and understanding; but doping is counterproductive to both the athlete's health and the image of the sport, so it is outlawed for medical and ethical reasons. Sports, however, have become progressively politicized over the years. National pride and prestige are at stake. Hitler knew that in the 1930s. Those who lived during the Cold War may remember the detailed, day-by-day and country-by-country medal counts.

Athletics have also become progressively monetized. Take, for example, Native American Jim Thorpe, whom the Associated Press ranked as the greatest athlete of the first half of the twentieth century. He won two Olympic gold medals in 1912, only to have them taken away because he had played two seasons of semiprofessional baseball, thus violating the rules at that time governing amateurism. No vestige of that high-minded purity remains today. In fact, money rules, with spiraling sponsorships, endorsements, and publicity.

Some athletes are naturally more endowed than others. Some have more opportunity. Some force themselves to train harder. Some think, "Wouldn't it be great to get just a teeny edge? Nobody will know." In some instances, athletes could (and likely do) dope out of season, reap the benefit of increased strength, and know, or at least hope, that no trace of the PED remains on board when it is time to get tested.

Doping issues extend beyond humans. They also involve horses, greyhounds, fox hounds, sled dogs, and even racing pigeons. As with humans, efforts to thwart these practices have had variable degrees of success.

If acceptable hormone levels in athletes have not been sufficiently complicated to date, new conundrums are emerging. These include determination of biological sex, which can be ambiguous, and consideration of whether rules should allow transgender individuals to compete with their naturally (or maybe unnaturally high but plausible) level of hormones. Let the games . . . go on.

Chapter 7

HUMAN CULTURE

NORMAL BODY CONTOURS, MOSTLY DEFINED BY MUSCLE, ARE aesthetically pleasing, and they probably account for the original and literal meaning of being in good shape. "Use it or lose it" is another expression likely used by various civilizations from early on. Socrates remarked, "No citizen has a right to be an amateur in the matter of physical training. . . . What a disgrace it is for a man to grow old without ever seeing the beauty and strength of which his body is capable."

Even long before Socrates, people seemed infatuated with physical training. Stone lifting was perhaps the first avocational form of weight training (after the vocational necessity of carrying the slain deer back to the cave). In China, stone lifting began as far back as 6000 BCE and was an element of religion, warfare, personal health, and custom.

From the time of early civilization, artists have sought interesting subjects and activities to record and have left a legacy that documents humans' interest in robust physiques and athletic activities. For instance, Egyptian tomb paintings from 3500 BCE depict men lifting sand-filled bags for exercise, and in various early civilizations artists sculpted leaders and gods shirtless and trim. In ancient Greece, trim turned to chiseled, and not surprisingly so, considering that no culture before or since has held society-wide health and physical fitness in such high regard. For the Greeks, physical and mental well-being were intertwined, as were their sports and arts.

Ancient civilizations portrayed their idols shirtless and fit.
Left: Egypt, c. 2480 BCE.
Center: Mesopotamia, c. 1800 BCE.
Right: Assyria, 640 BCE.

They held the first Olympic Games in 776 BCE to acknowledge these synergies. By around 600 BCE, Milo from the Greek city of Croton was weight training with animals. He was the most renowned wrestler of antiquity and was six times the Olympic champion. He purportedly gained his exceptional strength by carrying a calf daily until it became a fully grown ox. In due time, this exercise gave him the necessary strength to carry his own bronze statue to Olympia. One hundred years later, legend has it, Philippides ran the entire 26-mile distance from Marathon to Athens to announce a Greek victory over the attacking Persians.

Gymnasium comes from the Greek *gymnos*, meaning naked. (I am not sure which came first, nearly naked exercise enthusiasts or the gyms that condoned or encouraged such custom.) As part of their exercise routine, the Greeks hefted sculpted stones, about the size of a bread loaf, which had a through-and-through handle hole to accommodate a good grip. In addition to serving as handheld weights, long jumpers would swing a pair of these sculpted stones as

This Roman mosaic dates from the fourth century CE and comes from Villa Romana del Casale in Sicily. Various sports are depicted, along with the coronation of a winner in the lower panel.

they jumped. This shifted the momentum of the stones to their bodies and enabled the longest possible flight. The purpose of the stones might have been lost in antiquity if an artist had not documented their use on a Greek vase.

Artists likewise documented women's early participation in athletic competitions, which, for instance, are preserved in mosaics from the fourth century CE. Although such artistic renditions have proved to be priceless markers of history, they could not do justice to the definition of well-muscled figures because they lacked perspective and shading, which were Renaissance innovations. From the first, however, statuary was not so burdened. By the time of the ancient Greek civilization, sculptors' appreciation for balance and proportion—what we know today as "classical"—was, and is, in full view. Statuary, often displayed in public spaces, especially temples, had to be worthy of the gods. Romans co-opted these Greek ideals, reflected first in the muscularity of their gods sculpted in stone or bronze and not long after in representations of mortal warriors

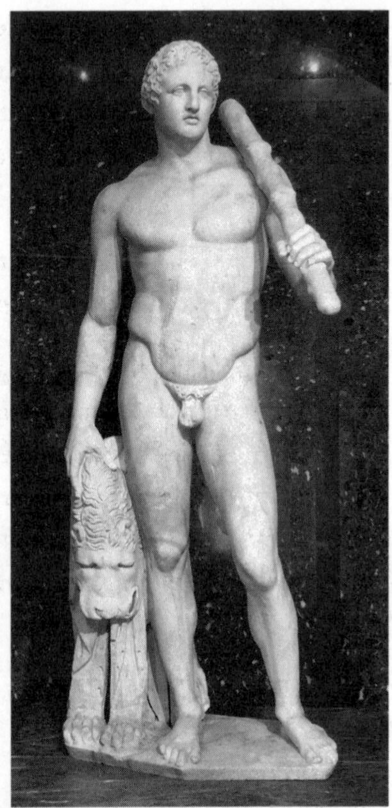

Sculptures from ancient Rome portrayed gods and heroes in robust form.
Upper left: Jupiter, Roman, 100 BCE.
Upper right: Heracles, Roman, 125 CE.
Bottom: Roman soldiers and Celtic warriors, 190 CE.

as well. Representation of muscles in these sculptures is remarkably lifelike, which is noteworthy, because it is highly unlikely that any of the early sculptors had ever directly observed human muscles, a disadvantage in realistic portrayals not to be overcome for another thousand years.

The Fitness Renaissance

Interest in physical fitness and artistic representation of buff bodies waned with the decline of the Greek and Roman civilizations. During the Middle Ages, modesty and humility took precedence over beauty or any appreciation of the ideal human form, so athletic activities short of warfare were generally disdained. This took a dramatic turn in the Renaissance, when interest in the Greek and

Illustrations from Girolamo Mercuriale's 1569 publication, De Arte Gymnastica, *show men lifting large tablets of lead (plumbum). Apparently women had not yet been allowed into the gym or to have spectators.*

Roman concepts of perfection was revived, and painters and sculptors studied artifacts to reestablish ideal body measurements.

Girolamo Mercuriale, a medical doctor who had access to the libraries of the Vatican and the families of Rome, studied physical conditioning in ancient texts and monuments and published *De Arte Gymnastica* in 1569. It was the first Western book on athletic activities. Mercuriale divided exercise into three groups: regular, military, and athletic. He prescribed ball games, walking, running, jumping, throwing, horseback riding, wrestling, boxing, and even dancing, singing, fishing, and hunting for both men and women. As in ancient Greece, mental and physical health worked in tandem.

Intertwined with the new focus on art, anatomy, and engineering that began in the Renaissance was a return of interest in physical culture. The practice of pumping stone in ancient China and sand in ancient Egypt reemerged as pumping lead in Renaissance Italy and eventually evolved into pumping iron. The original dumbbells appeared in the 1700s, when some blacksmith placed a church bell on each end of a rod and removed the clappers, rendering the bells silent—or *dumb*. Ben Franklin, an early aficionado, wrote in 1786, "I live temperately, drink no wine, and use daily the exercise of the dumbbell." Gravity being what it is, people have always been dropping free weights on their toes, which gave rise to weight machines, one of which appeared in 1796, just ten years after Franklin extolled the virtues of dumbbells.

In the nineteenth century, American Catharine Beecher, sister of Harriet Beecher Stowe, advocated strongly for women's education and promoted calisthenics as a means of physical conditioning. Various images of gymnasiums from this time give evidence of the growing interest in physical fitness—for both sexes, separately—and for the wearing of more clothes than did Mercuriale's subjects from the sixteenth century. Fencing equipment, Indian clubs, parallel bars, high bars, vaulting horses, ropes, ladders, poles, weights, and spectators abounded. In another image of the time, a woman, modestly clad in bloomers and a loose top, is executing a perfect iron

Fencing, climbing, using the high bar, vaulting horse, trapeze, balance beam, and spectating were among the activities available at Roper's Gymnasium in the 1830s in Philadelphia.

cross on the rings. You go, girl! Women first participated in Olympic gymnastics in 1928 and in weight lifting in 2000, whereas these sports for men were part of the first modern-era Olympic Games, which took place in 1896.

Eugen Sandow

Concurrent with the modern Olympic movement, the recurring emphasis on the perfect human physique came to captivate Prussian Eugen Sandow (1867–1925). He described his early teenage self as weedy, fragile, and pale-skinned. When he was 15, his father took him on a trip to Rome, where the young man became infatuated with the brawny brutes represented in Greek and Roman statuary. He dreamed of emulating them, and on return home Sandow committed himself to achieving a state of physical perfection. To determine exactly what the "Grecian ideal" should be, he visited museums and measured statues' various dimensions. Then, taking

Eugen Sandow (1867–1925) stressed the development of muscle size and definition and became the father of modern bodybuilding.

tips from circus strongmen, Sandow began pumping iron to achieve the Grecian ideal of broad shoulders, tapered back, small waist, and detailed but overwhelmingly huge muscles. He performed lots of reps. It worked (and it still does).

Sandow stressed the aesthetics of muscle size and definition over strength, but he became strong in the process. At 18 he left home and toured Europe first as a circus athlete and professional wrestler and later as a weight lifter. Eventually, Florenz Ziegfeld contracted him to show off in Chicago at the 1893 World's Columbian Exposition. Once Ziegfeld discovered that the audiences were more interested in Sandow's robust physique than in how much weight he was lifting, he asked Sandow to pose and flex in what he called "muscle display performances" to highlight various muscle groups.

Subsequently, Sandow traveled the world, published *Sandow's Magazine of Physical Culture*, and name-branded a line of cigars, possibly the first celebrity endorsement of a commercial product. In his performances, Sandow dazzled audiences with feats of strength such as lifting pianos, bending iron bars, bench-pressing cows, and tearing decks of cards in half, a stunt for which he was once bested. A young man in the audience, who later became the "World's Strongest Youth," jumped on the stage, took half of the torn deck from Sandow, and tore it in half again.

In 1901, Sandow hosted the first major bodybuilding show, "The Great Competition," for an overflowing crowd at the Royal Albert Hall in London. Clad in leopard-skin leotards, the 60 contestants flexed their muscles for the enthusiastic crowd and the discerning eyes of the three judges—Charles Lawes, a sculptor and athlete; Arthur Conan Doyle, author and friend of Sandow; and Sandow himself. The winner received a gold-plated statuette of the competition's nearly naked host.

For all these accomplishments, Sandow is known as the father of modern bodybuilding. (Take note: bodybuilding contests stress form, not strength. By contrast, powerlifters and Olympic weight lifters are not judged at all on their appearance, just on their strength and explosive speed, respectively.) Sandow inspired the golden age of bodybuilders, which included Steve Reeves, Frank Zane, and Arnold Schwarzenegger. Today, the award for the Mr. Olympia contest is called "The Sandow."

Charles Atlas

Further impetus for the coming physical fitness craze came from a 97-pound weakling born 25 years after Sandow, in 1892. We better remember Angelo Siciliano as Charles Atlas. As described in his promotional brochures, Atlas decided to improve his physique after a bully kicked sand in his face at the beach. Initially humiliated, Atlas resolved to bulk up. It was perhaps his lack of ready access to a gym or weights that caused an epiphany at the zoo one day when he watched a lion stretch. The lion used just the resistance of its body weight to stay fit. This led Atlas to develop, use, and eventually market his regimen of "Dynamic Tension," which was mainly isometric resistance exercises. Atlas's method used one muscle group to resist another through a range of motion without using weights. For example, he would sit in a chair and raise his knee while pressing against it with his hand to strengthen his hip flexors and elbow extensors.

Atlas's bulking up coincided with a boom in the placement of public sculpture, and he modeled his "knockout" body for various artists, including Alexander Stirling Calder (father of artist Alexander Calder). Among the many delights of Washington Square Park in New York's Greenwich Village is the sight of Charles Atlas dressed up as George Washington.

After that, Atlas's career as a model, champion bodybuilder, businessman, and showman took off. He won a photo contest in 1921 for "World's Most Beautiful Man." This motivated the same sponsor to host a live "World's Most Perfectly Developed Man" contest the following year. Judged by a panel of doctors and artists, Atlas won over 774 competitors. The sponsor, certain that Atlas would win every year, declined to host any more contests. Although a modest and decent man, Atlas took advantage of opportunities for self-promotion. Numerous photographs show him hefting bathing beauties, playing tug-of-war against the Radio City Music Hall Rockettes, and joking around with professional boxers.

Atlas marketed his Dynamic Tension pamphlet on the back of comic books. The advertisement featured cartoon frames of a skinny beachgoer returning months later, jacked, to intimidate the bully. An obviously impressed girl exclaims, "You're a he-man now!" Under the headline "I can make YOU a new man, too, in only fifteen minutes a day," Atlas included a nearly naked photo of himself. It was a marketing masterpiece—a pamphlet was much easier to print and mail than exercise equipment. The pamphlet was eventually translated into seven languages and in the 1950s was responsible for drawing 40,000 new Dynamic Tension recruits annually.

Atlas probably also pumped iron. Most experts agree that isometric exercises alone will not suffice, and progress is easy to measure when lifting weights. Isometrics do, however, present a far lower risk of injury.

Sandow and Atlas were similar in several ways. Both were committed to muscle building and were unapologetic self-promoters. As the twentieth century progressed, however, these men's legacies took different paths. Since Sandow was born 25 years earlier and died 47 years before Atlas, I will describe his enduring influence first—bodybuilding with emphasis on form rather than strength.

Bodybuilding

It is probably not entirely coincidental that synthetic testosterone became available in 1935 and that the Mr. America bodybuilding contest started in 1939, Mr. Universe in 1948, and Mr. Olympia in 1955.

Female bodybuilding contests trailed several decades behind male ones, and, as one might expect, audiences differed on what form a female winner should take—some resemblance to a Grecian goddess's physique or just hugely muscular. The best of both are rewarded, because today women bodybuilders can compete in five categories: Physique, Figure, Fitness, Bikini, and Wellness. Universally evaluated are poses in increments of 90-degree turns along with demonstrated poise throughout the judging. The fitness category also requires a dance routine that includes a push-up, high

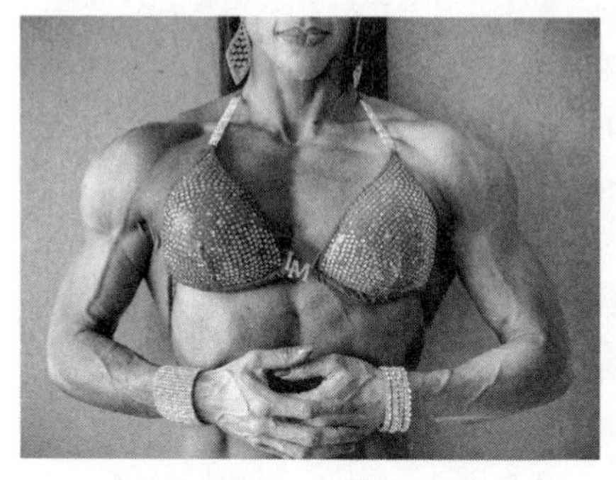

Above: A buff female fitness figure competitor shows the results of her devotion to dieting and weight lifting. Right: Seven-times voted Mr. Olympia, Arnold Schwarzenegger performs the "front double biceps" pose.

kick, straddle hold, and side split. Some categories call for high heels and jewelry, others for bare feet.

To prepare for a competition, male bodybuilders tell me, it takes about three months of disciplined work to go from routinely robust to awesome. They continue with their usual strenuous weight-lifting routine and begin intense endurance training and dieting to reduce their body fat, which masks their musculature. Their diet during the final month consists mostly of chicken breast, fish, broccoli, asparagus, and a little rice supplemented with nutrients and protein shakes, which in total might add up to 1,000 calories per day. The aim is to reduce body fat to 3–7 percent of the total body, which from a general health perspective is dangerously low. Several days before the competition they begin loading up on carbohydrates to add glycogen and bulk to their already bulging muscles. For the last 36 hours they restrict water and sodium intake, leaving themselves starved, dehydrated, and likely irritable, but now nothing but skin separates their muscles from each judge's scrutiny. Additionally, some bodybuilders indulge in a pre-performance belly-flattening enema. Because of the overall commitment required, most bodybuilders do not compete more than once or twice a year.

For the event itself, a natural tan will not hold up under the bright lights, and advice on the Internet includes, "When you think you're tan enough, do another two coats! Judges can and will hold a poor tan against you, so err on the side of caution and assume that more is better." To further take advantage of the bright lights to demonstrate muscle definition, bodybuilders shave with safety razors. Electric ones don't trim close enough.

For the women, the bikinis are small, sparkly, and require particular attention to detail. "Since you need to make sure you are 'secured' in your suits, make sure to bring some suit glue."

Male bodybuilders compete in three categories: Men's Physique, Classical Physique, and Men's Bodybuilding. The typical body shape/size by category is robust, plausible, and unbelievably freakish, respectively. As the muscles get larger from one category to the next, the outfits get smaller: board shorts to small briefs to even smaller "posing briefs." In Men's Physique, the contestants are judged from both front and back but without any blatant muscle flexing. For the Classic Physique and Bodybuilding categories, each contestant assumes eight different required poses. These include front and rear double biceps, front and rear lat (latissimus dorsi) spreads, side triceps, side chest, and front abs and thighs. Following the obligatory poses, contestants have an opportunity during the "pose down" to flex freely in an individualized, choreographed sequence of postures that they think maximally display their splendors. In case you think bodybuilding contests have passed you by, some of them have age categories to accommodate everyone, including masters (over 39 years old), grand masters (over 49), and even super ultra platinum masters (over 79).

For both men and women bodybuilders who, despite overtime in the gym, lack symmetry or who can't make a particular muscle stand out, assistance is available via Synthol injections or surgical implants. Just as the use of growth-enhancing medicinals such as steroids, growth hormone, and insulin are unregulated in most competitions, so are physical-bulk enhancers. Synthol, which is mostly oil with some local anesthetic and alcohol mixed in, is advertised as a posing oil with characteristics purportedly better than baby oil and olive oil. Some bodybuilders, however, use it as a "site enhancement oil" and inject it to "fluff" out an otherwise perfect physique. (Why else would it contain local anesthetic?)

To learn more about this performance art, I recently attended a bodybuilding competition—as a spectator. The International Natural Bodybuilding Association was the host. It takes great pride in being one of the bodybuilding organizations that pays more than lip service to prohibiting use of performance-enhancing drugs. The INBA randomly picks contestants for testing and routinely tests winners in each category.

I paid extra for a backstage pass to visit the prep rooms, which early on had standing room only and displayed vast expanses of "tanned" skin tightly stretched over bulging muscles, none freakishly large in accordance with the INBA's firm stance against performance-enhancing drugs. Contestants not satisfied with the sheen provided by their newly applied spray tan were applying posing oil to themselves and to the backs of fellow competitors. Some had brought their dumbbells along and were "pumping up" their muscles into full glory. Others were doing slow, controlled push-ups between strewn-about gym bags laden with supplements. I glanced into the women's ready room but immediately turned away in shock. Amazons! In bikinis and high heels! With eye shadow!

The audience nearly filled the 300-seat theater, and everyone I talked to was either a friend or family member of one of the competitors. Seven judges, male and female, all former bodybuilders, sat in the front row with clipboards in hand. Several rows back a professional photographer clicked away all day, providing images for the INBA's magazine, *Ironman*. A forest of trophies covered five or six tables at the back of the stage. For each category, the emcee introduced competitors by name, age, city, time in training, and day job. Many were personal trainers, but the clergy, police force, and business interests were also represented. In categories that had at least three competitors, the host awarded checks of $1,000, $500, and $300 to the top three winners. In addition, the trophy tables were gradually deforested over the day.

For part of the show (after the Amazon jolt, as I recall), I sat next to a woman whose husband was competing. They had flown into Los Angeles from Austin, Texas, the day before, were staying in a hotel for two nights, and then flying home. He won his category, which included seven or eight competitors, so naturally his wife excitedly photographed him holding up his giant $1,000 check. Some silent calculations convinced me, however, that by the time they got home, they would be ahead by only the trophy and maybe some brag rights at their gym—"their gym" because sometimes she also competes. Maybe his photo in *Ironman* would garner him a product endorsement, movie audition, or additional clients to train, but it had to be more than prize money that motivated him.

I had an even more depressive thought as the day progressed.

After they had performed, showered off their tans, and donned warm-ups, some of the contestants came out and sat in the audience. Although they looked trim and fit, their clothing completely disguised their lean physiques and awesomely developed musculature. For instance, there was no evidence whatsoever that their abs looked like biscuits on a baking sheet and that their silhouettes when performing their front-lat spreads reminded one of a hulking B-52 bomber. Consider that the men rigidly diet for at least three months to reduce their body fat to 5 percent of their overall weight. (The American Council on Exercise says that "fit" and "athletic" men have about 16 percent and 9 percent body fat, respectively. For women, the averages are several percentage points higher.) Then the contestants pump iron obsessively and especially spot-train muscles that aren't quite as grand as their others to gain "symmetry." After all of this they might win a trophy and a break-even weekend in Southern California.

Fitness Today

Whereas Sandow is appropriately known as the father of modern bodybuilding, Atlas, perhaps because he came along later and lived longer, has had a far greater societal influence on trends in physical conditioning. He pioneered the modern fitness boom and helped introduce the era of youth consciousness. Following in Atlas's footsteps were Jack LaLanne, Victor Tanny, Joe Gold, Richard Simmons, and Jane Fonda. They opened gyms or promoted athletic fitness via mass media. More Americans began exercising regularly, jumping from 24 percent in 1960 to 69 percent in 1987.

Dr. Kenneth Cooper, widely credited with being the father of the modern fitness movement, coined the word *aerobics* and published his groundbreaking book on the topic in 1968. He gave ordinary people license to put on gym shorts and jog through the neighborhood without shame.

Two seminal events occurred in 1972, the year of Atlas's death.

Dandies Dressing *by Robert Cruikshank, 1818. Captions: Man in top hat: "Tom you are a charming figure! You'll captivate the girls to a nicety!!" Man with back turned: "Do you think so Charles?—I shall look more the thing when I get my other calf on."*

The first Nautilus fitness machine came on the market, and the US Congress passed Title IX, which stated, "No person in the United States shall, based on sex, be excluded from participation in, be denied the benefits of, or be subjected to discrimination under any education program or activity receiving Federal financial assistance." Women's school-based athletic programs blossomed, yet some would bemoan that this shift sometimes caused reduction in funding for men's college sports.

Barring the time or inclination to pump iron, what can the average Jack or Jill do to look buff? Fake muscles have been available for over 200 years. In *Typee*, published in 1846, Herman Melville wrote, "Stripped of the cunning artifices of the tailor . . . what a sorry set of round-shouldered, spindle-shanked, crane-necked varlets would civilised men appear! Stuffed calves, padded breasts, and scientifically cut pantaloons would then avail them nothing, and the effect would be truly deplorable." Favorite padding materials for the time were sawdust

Maureen O'Hara, 1947, wears a stylish jacket of the time with squared shoulders.

and paper. The Internet now offers "silicone calf pads for crooked or thin legs," but one user bemoans, "Can someone post a pic on how u do it? I use heavy sock but it doesn't stay up." Bulgy calves are apparently as irksome to some individuals as skinny ones are to others. For the bulgers, Botox injections, liposuction, and surgical partial denervation of the calf muscles all show appealing results on the Internet. The sparse literature on the subject does not describe the owners' ability following treatment to stand on their toes or to run. What price beauty?

Padded undergarments with sewn-in shoulder, chest, thigh, and tush enhancers are also available. Remember, too, that off and on for

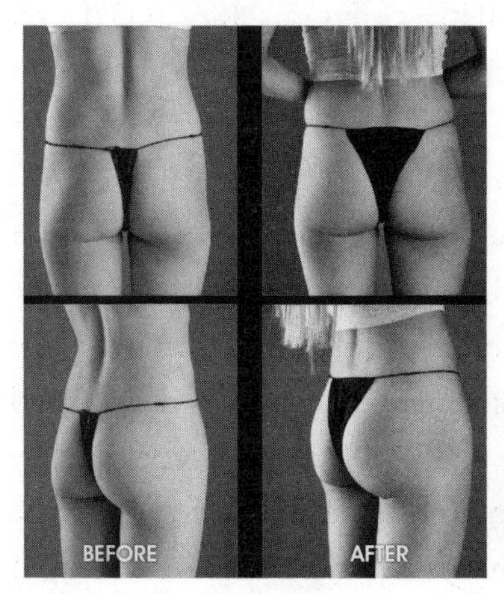

Left: Before and after silicone gluteal implants.
Below: Before and after a right silicone biceps implant.

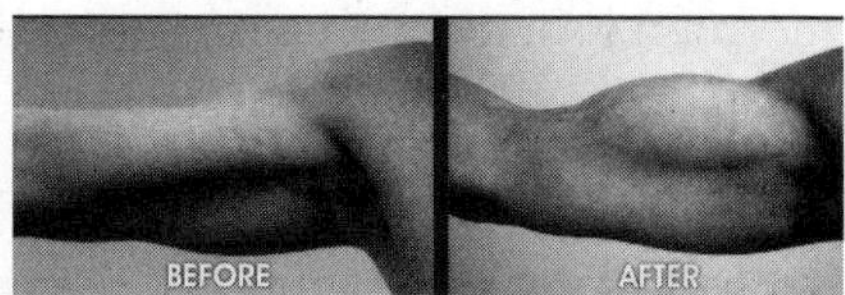

nearly a century fashionable women have worn clothing that emphasizes large, squared shoulders and high-heeled shoes that draw attention to the calves and tush. (Gentlemen, before getting too smug, remember that men's jackets at times have sported concealed padded shoulders. And what effect do you think epaulets create?) Recently bare midriffs (for women only, thank goodness) have been in vogue, and a six- pack can be suggested with strategically applied cosmetics.

What about well-heeled Jacks and Jills who desire a robust appearance even when they're barely clothed? Cosmetic surgeons can bulk up their guns (bro-speak for biceps), delts, traps, lats, pecs, glutes, and calves with silicone implants. Either implants or selective liposuction offer six-pack enhancement. Silicone implants are also image-restorers after an accident or when tumor surgery has vastly damaged a normal body contour.

Obviously, muscles are on people's minds. We like having them and we like seeing them. They radiate vigor and vitality presently as they did in ancient Greece, and now they not only pervade fine art but also popular culture. Popeye proved that vegetarianism was no block to building muscle. Rosie the Riveter put on a gun show and inspired women to join the war effort in the 1940s. She sub-

The poster We Can Do It! *(better known as Rosie the Riveter) by J. Howard Miller lifted American workers' morale during World War II.*

sequently became a symbol of American feminism. Then consider Wonder Woman, Superman, Black Panther, Green Lantern, Thor, King Kong, The Incredible Hulk, He-Man. By comparison, big hearts and big brains just cannot compete in our visual imagination with superheroes' bulging muscles.

Also deep in our collective psyche are Muscle Beach, where the modern fitness boom in California began in 1934, and muscle cars—street-worthy dragsters with engines on automotive equivalents of testosterone. Colorful expressions describe weight lifters' and bodybuilders' developmental accomplishments. "Cut" and "shredded" describe entry-level muscular development and require good definition of abs and upper-extremity muscles, but none of them need to be particularly large. "Ripped" comes next and implies impressively sized musculature. "Jacked" and "swole" and "yoked" are reserved for breathtakingly large physiques, perhaps grotesquely so.

Other less well-defined descriptions include beach bod, beast, buff, built, and diesel. If you want to up your cred as a gym rat, you should also be ready to acknowledge a Beef Swellington, Swoldier, Brodozer, Hard Body, or Masster when you see one achieving a personal best.

I suspect humans have been noting personal bests for as long as

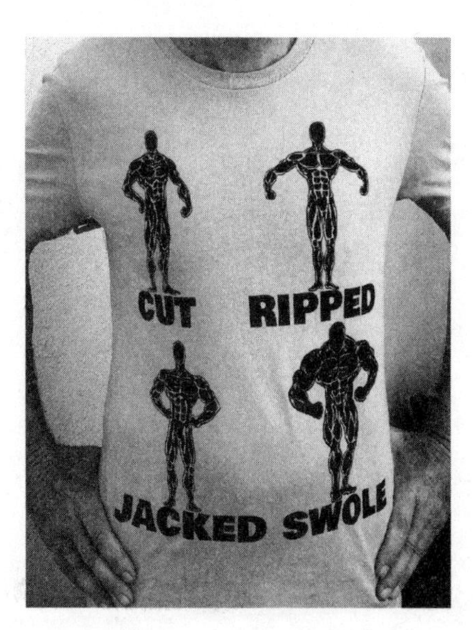

"Bro" definitions of bodybuilders' physiques.

there have been humans, competing first with others as a matter of survival, later against themselves for ego satisfaction. It's likely that the Chinese stone lifters in 6000 BCE and the Egyptian sandbag slingers 2,500 years later made contests out of those endeavors and sought bragging rights. Muscles can certainly demonstrate stunning feats of speed, agility, strength, coordination, and endurance. Even under less stressful activity or with no obvious insult, muscle disorders do occur. That should not be surprising considering that muscles range from head to foot and account for 40 percent of our composition.

Chapter 8

DISCOMFORTS AND DISEASES

SKELETAL MUSCLE MALADIES (UNLIKE SMOOTH-MUSCLE AND cardiac muscle maladies) deserve their own chapter for two reasons. Skeletal muscle constitutes a far greater proportion of our body mass than the other two types combined. So by size alone, it captures our attention, particularly since it is immediately beneath our skin and is more directly observable than smooth and cardiac muscle. Then, because skeletal muscle is mostly under our voluntary control, we test its state of health with every movement, learn quickly when something is amiss, and seek the causes and remedies for even minor maladies.

Considering skeletal muscle's high level of energy expenditure, longevity, resilience after abuse, and adaptability to its owner's wishes, seriously bad things happen to it only rarely. Even when a joint isn't moving properly, it may not be the muscle's fault, but rather the result of a faulty message originating in the brain, spinal cord, peripheral nerve, or neuromuscular junction. Skeletal muscles cross one or more joints, and if a joint is stiff or fused from any one of several disorders, the best muscle contraction is not going to achieve the desired movement.

Whole-body woes that affect skeletal muscle include wasting, some of which naturally occurs with aging. One cause of earlier onset is a lack of adequate resistance exercise. This may result from inactivity or, in recent times, from spending weeks in a low-gravity environment. Other causes include inadequate nutrition and hormonal derangements. For example, faulty levels of growth hormone, insulin, and testosterone have detrimental effects on muscle health.

Cerebral palsy, stroke, tremor, Parkinsonism, and Huntington's disease are well-known examples of problems originating in the brain that manifest themselves as movement disorders.

Farther along the control chain are spinal cord conditions including injury, amyotrophic lateral sclerosis (Lou Gehrig's disease), and, historically, polio. They adversely affect the muscles' nerve supply and result in wasting.

Even farther along the control chain, the nerves extending from the spinal cord to the muscles can sustain injury by laceration or compression (for example, carpal tunnel syndrome) and from more exotic conditions including Guillain-Barré syndrome, Charcot-Marie-Tooth disease, and chemical toxicity (DDT, for instance). The neuromuscular junction is where the electrical impulses arriving at a nerve's end stimulate chemical messengers (neurotransmitters) that relay the message into the muscle. Two types of problems here ultimately affect the muscle—either too much neurotransmitter, causing overstimulation of muscle contraction, or not enough transmitter, causing paralysis. Causes for muscle hyperactivity and involuntary contractions include bacterial toxicity (tetanus, also known as lockjaw), pesticide poisoning (malathion, diazinon, lindane, strychnine), and nerve agents (sarin, VX, Novichok). Conditions characterized by inadequate production of neurotransmitters or blockage of their effect and causing paralysis include the disease myasthenia gravis. Botulinum, curare, hemlock, atropine, and venom from kraits, mambas, cobras, and coral snakes are examples of paralyzing toxins.

Fractures, dislocations, arthritis, and infections may stiffen joints and prevent movement even in face of entirely normal muscles. The muscles contract isometrically but no motion ensues.

✦✦✦✦

At the end of the control chain, skeletal muscle itself can become disordered, and the ways it can do so are the focus of this chapter. You likely have had some of these maladies—such as fatigue, cramps, soreness. This book is no substitute for a medical encyclopedia or your doctor. Rather I highlight a sampling of these conditions to enhance your understanding of some common ones and to introduce you to a couple of rare ones. Together they illustrate skeletal muscle's complexity and its exceptional routine durability. Although much is known about muscle's function and dysfunction on molecular and cellular levels, major mysteries remain unsolved. Let's have a look.

Unexplained Aches and Pains

The only thing known for certain about growing pains is that the condition has been misnamed for 200 years. A French physician first described it in his 1823 book, *Maladies de la Croissance* (Diseases of Growth). Perhaps as many as 40 percent of four-to-eight-year-old children awaken at night once or twice a week with pains in both calves and thighs that may last several hours. The episodes are not associated with growth spurts, so some sticklers would prefer a more accurate name, such as benign idiopathic paroxysmal nocturnal limb pains of childhood, but even that descriptive mouthful doesn't explain the cause. Muscle fatigue after the child has been particularly active may play a part. Vitamin D deficiency, bone density, circulation, psychological factors such as stress in the family, and a lower-than-normal pain tolerance are among the possible contributors that have been explored; but none have been definitively linked. The good news is that the aches typically benefit from massage and heat, they are benign, and they eventually disappear. Hence it is not a big public health issue and will likely remain unfunded by major research-granting agencies.

I wish, however, that somebody would fund research to find a cure for morning stiffness, which I notice just as I get out of bed every morning, especially if I exercised strenuously the day before. I probably look like a drunk for my first couple of steps, not sure of my balance. My muscles don't quite want to do what my brain is asking of them. The muscles seem tight, maybe achy, not sore. Then with a couple of minutes of warm-up exercises (swinging arms while marching in place, neck rolls, torso twists), the stiffness disappears. I then notice it again later in my hips and knees if I stand up after sitting for a couple of hours. On reflection, I would notice stiffness decades ago when first getting out of a car after a long trip. I groan and hobble a bit, then it's gone. My patients notice it when they first come out of a cast that had been immobilizing a major joint.

Some people, especially after surfing the Internet, determine that rheumatoid arthritis, an autoimmune inflammation of the joints, is the cause of their joint stiffness. Yes, morning stiffness is a classic finding in rheumatoid arthritis, but the diagnosis also hinges on some combination of swollen, hot, tender joints that are painful to move and blood tests that are usually positive for inflammatory markers. Hence, nearly all instances of isolated morning stiffness have nothing to do with rheumatoid arthritis. Whew!

On the one hand, I have not been able to find any scientific reports on garden-variety morning stiffness, especially when scientists can explain the molecular structure of myosin, the jumping mechanism of fleas, and the function of salamander tongues in freezing weather. On the other hand, it is reasonable to ignore and not study such an isolated symptom for several reasons. Morning stiffness goes away on its own, it's not fatal or even debilitating, and I am not sure how anybody would go about studying it, even if some deep pocket were willing to come up with research money. I for one would not be willing to have a muscle biopsy when I first get up. My best guess is that water, which bathes all our cells, tends to settle where it wishes when our muscles are still, and then it's in the way until we start moving and squeeze it around a bit. Call it a gel phenomenon.

Stiffness with cold exposure is easier to explain. When there is not enough heat to go around, the sympathetic nervous system favors keeping the internal organs warm at the expense of the limbs.

People often note generalized muscle achiness associated with having the flu or having received an immunization. No more is known about this symptom than about morning stiffness or growing pains. All three of these conditions occur without any apparent use or abuse of skeletal muscle.

Activity's Aggravations

Physical activity and overactivity, however, do contribute to the following common conditions: muscle strain, delayed-onset muscle soreness, cramps, and fatigue.

A muscle strain stretches the tissue beyond its natural limits. (By contrast, a sprain results when a ligament, spanning bone-to-bone across a joint, sustains such a stretch.) The severity of the strain varies according to the degree of stretching. Think about donning a tight pullover sweater. Sticking your head through and then slowly inserting one arm at a time, you might hear a few threads pop, but the sweater keeps its shape. Going faster pops more threads. Trying to get your head and both arms quickly positioned, something rips, and the sweater is forever misshapen. That is why a strained muscle is accurately called a pulled muscle. While a muscle is contracting, a yank that suddenly lengthens it will cause something to give way. It may be the tendon's attachment to bone, or it may be in the substance of the tendon or muscle itself or at the junction where the muscle transitions into tendon. Sharp pain, localized bruising from torn blood vessels, and swelling ensue. Thereafter, any contraction of the muscle produces pain.

The hamstring muscles are particularly prone to strain injuries, which are typically sustained while running, especially when it includes sudden starts and stops. The three hamstring muscles originate underneath the gluteus maximus on the pelvis, cover the back of the thigh, and insert just beyond the knee. Thus, on contraction they straighten the hip and bend the knee. In isolation, these motions would occur if you kicked a ball backward using your heel. In concert, these motions occur with every walking step. Even at ambling speed, the hamstrings must contract eccentrically and then a fraction of a second later contract concentrically before instantly relaxing. They repeat the cycle as the hip and knee joints move back and forth in a complex and carefully synchronized manner. Hence, basketball, football, and tennis maneuvers may ask the hamstrings to do too much at once and result in a strain.

The low back muscles are another frequent location of strain injuries, often described as "I threw my back out." Squatting and lifting heavy objects with the legs rather than bending over and lifting with the back spares these vulnerable back muscles. Use of a wide abdominal support belt favored by weight lifters and warehouse workers further protects the low back muscles. During the lift, the "corset" supplements core muscle strength, keeps the low back flat, and relieves some pressure on the spine. After the lift, the belt should be released immediately, otherwise it shields the core muscles from the training they need to become and remain strong.

For any athletic endeavor, warming up the susceptible muscles slowly and fully minimizes the risk of pulling one.

The mnemonic RICE stands for rest, ice, compression, and elevation and describes the advocated and well-worn treatment for sprains and strains. Rest is easy and intuitive, because a tear hurts. Intermittent icing may reduce pain. Compression in the form of an elastic wrap might hasten the resolution of swelling for a muscle strain at the knee or ankle where it is anatomically feasible, but it is impractical for the low back or for a strain of the hamstrings from their origins on the pelvis, which, if you are sitting, is where your body is putting pressure on the chair right now. Again, elevation works for knees and ankles because getting the injury higher than the heart allows the veins to drain and helps relieve swelling. Whoever recommended RICE for low back and hamstring strains, however, really hadn't thought through what gymnastics would be required for a person to elevate those parts higher than the heart and then try to rest.

Overall, RICE is the best we have for strains, but it basically comes down to R—rest. A tear heals with tough but noncontractable scar tissue, but depending on the exact location and extent of the strain, recovery may take weeks and is sometimes permanently incomplete. Injecting platelet-rich plasma (PRP) is a currently popular means to hasten healing of muscle strains, but laboratory support of its effectiveness is mixed, and large-scale clinical trials are lacking.

The thinking behind PRP is that platelets are loaded with various protein messengers—growth factors—that promote cell growth and healing, two cell processes that are immensely complicated. Consider an analogy where growth factors are musical instruments. In nature, the interplay of growth factors is part of a beautiful symphonic performance, with the violins and woodwinds coming in on cue as the trombones fade away, for example. An injection of PRP, however, dumps a truckload of musical instruments on the street and expects them somehow to sound musical. Perhaps one day investigators can modulate the concentrations, appearances, interactions, and disappearances of growth factors in a way that is truly symphonic. Presently, however, headlines sometimes report that a high-profile athlete received PRP injections and is now back in the game. What the fine print misses is whether the PRP was truly responsible for the healing process or if it was coincidental to RICE and the result of natural healing.

How about the pain one gets one to three days after participating in some unaccustomed activity, especially one that eccentrically loads a muscle group? For instance, a headbanger at a heavy metal concert may awake the next day with a sore neck.

The eccentric-loading activity that affects me is hiking down a mountain all afternoon. I don't do it often, and the required ankle motion requires contracting my calf muscles while gravity and body weight are making them longer. I am fine until the next day, when my calves hurt so badly that I descend stairs by hanging onto the rail with both hands or by easing myself down backward. Remarkably, though, my calves are not particularly sore to pressure, nor are they sore when I am just sitting, standing, or walking on level ground. Welcome to delayed-onset muscle soreness. Those who think they understand the condition may toss around the acronym DOMS. They don't. Nobody does for certain.

Delayed-onset muscle soreness most likely results from structural disturbances in the arrangement of the actin-myosin units in the overworked muscle. This leads to shifts of fluid, inflammatory mediators, and calcium and sodium ions in and out of the cells. Recent investigation has also implicated the thin overlying membrane of fibrous connective tissue surrounding the muscle units. Despite active research using precise investigative methods that include magnetic resonance imaging and biopsies examined with electron microscopy, the complete answer is elusive. Lactic acid buildup used to be the knee-jerk explanation, but experts have thoroughly discounted this idea because any accumulation disappears within an hour after strenuous exercise. In addition, all types of exercise produce lactic acid, and if it were the culprit, then DOMS should occur whether the exercised muscles were contracting eccentrically or concentrically. Hiking or running downhill forces the calf muscles to repeatedly contract eccentrically, and such activities are notorious for causing DOMS. Although hiking and running uphill are more strenuous, the calf muscles contract concentrically and produce lactic acid, but going uphill does not typically result in DOMS. For me, trying to decipher the cumulative research is like hearing from the blind men feeling an elephant. Each investigation identifies interesting findings, but in summation the information does not fit together.

If the cause of DOMS is cloudy, so is its treatment. Anecdotal information is freely exchanged and likely worth what the recipient pays for it. Good studies often produce conflicting results, possibly because of variations in subjects studied and methods of analysis. Gentle foam-roller massage and light exercise seem to help, at least improving comfort but not necessarily hastening performance recovery. Heat probably aids circulation to the sore area and may help resolve inflammation and restore normal muscle metabolism.

Athletes also jump into ice baths after strenuous exercise, and the shock value of a photo taken while sitting chest deep in ice cubes may garner desired attention on social media. If ice is not cold enough for you, there are also whole-body, walk-in cryotherapy units that expose the body to as low as –220 degrees Fahrenheit for several minutes. The efficacy of icing in any form, however, is unproven, and any claims of benefit may be no more than a placebo effect. Proponents may fool themselves by thinking, "Something this uncomfortable must be doing me some good." The inability to blind subjects from knowing whether the tub of water is at room temperature or ice cold blunts any effort to investigate the effects of icing scientifically. Yes, the cold does cause the blood vessels to constrict, and proponents claim that icing reduces post-exercise inflammation, but they overlook the fact that inflammation is a necessary part of the muscles' adaptation to the recently completed training session. So icing is not a rational treatment after strength training if the goal is to get stronger. It might, however, temporarily reduce pain and thereby allow athletes who are seeking short-term recovery between events, and who are not concerned about long-term adaptations, to perform better.

The value of using herbal preparations for prevention of or relief from DOMS is equally murky. Because of their antioxidant properties, black currants, lemon verbena, tart cherry juice, saffron, and curcumin have their advocates but, again, good scientific support is lacking. Time, however, heals all wounds, and once the DOMS-affected muscle recovers, it is resistant to suffering a repeat bout, at least for a while.

Cramps, sometimes exercise-induced, are another enigma. These temporary, involuntary, painful contractions are poorly understood and hard to study, partly because they are unpredictable and transient, and because for most people they are merely irksome, not disabling. When one occurs in the legs, it is a charley horse. (This expression seems to have come from professional baseball in the late 1800s when it suddenly began appearing in newspaper articles. Nobody is certain how the expression came about, perhaps from the athlete's limp, which was reminiscent of a rocking horse.)

Cramps that may come on with exercise, or soon after, are possibly related to a water or salt imbalance in some cases and perhaps start with a spinal cord reflex rather than within the muscle itself. A slow stretch of the tightly contracted muscle along with massage seems to help, especially if there is time to get the offending part under warm water. For me, grimacing seems obligatory, although I am not sure whether that is part of the problem or part of the solution.

Doctors used to prescribe quinine as a preventive. It was effective, possibly by decreasing conversion of electrical signals from the nerve into chemical signals in the muscle and possibly by increasing the brief but obligatory intervals between muscle contractions. Taking quinine for cramps, however, came with unacceptable side effects, including irregular heartbeats and platelet depletion.

Is fluid imbalance the cause of cramps? In one study, subjects ran downhill on a treadmill (eccentrically contracting their calf muscles) in a chamber heated to 97 degrees Fahrenheit while they hydrated either with water or with a sports drink containing glucose and markedly more sodium, potassium, and magnesium than the plain water. Beginning immediately after a 3- to 5-mile simulated run and again 30 and 65 minutes later, the researchers shocked these volunteers' calf muscles with progressively increasing electrical jolts until cramping occurred. At each time interval, the electrical voltage necessary to induce cramping was significantly higher in those who had consumed the sports drink. In other words, it reduced the tendency to cramp.

Another set of enterprising scientists studied the effect of pickle juice on cramps. College students (more volunteers!) agreed to have the nerve to the muscle that curled their big toe down stimulated with a mild electrical shock. The stimulus was applied repeatedly until the fatigue generated cramping (just as treadmilling downhill in the heat had caused fatigue-induced cramping in the other study). Then the toe-cramped subjects swallowed 2.5 ounces of either pickle juice or water, the exact volume adjusted by each subject's body weight. (To prevent visual or olfactory clues from clouding the results, the liquid came in an opaque container, and both the researcher and each subject wore nose plugs.) Cramp duration was 88 seconds after swallowing pickle juice compared to 134 seconds after ingesting water. The study leaders, however, could not explain the effect by a rapid restoration of body fluids or salts because the relief occurred so quickly. Rather, they could only conclude, by exclusion, that a reflex in the throat probably inhibited the irritable nerve responsible for the cramping.

(Before I had thoroughly read the study, my computer-mousing hand cramped one day, so I headed for the fridge for a swill of pickle juice. It had no effect. Later I learned that the study subjects tossed back a considerably larger volume. I haven't had another cramp to repeat the test. It may be that my body is holding off on cramping knowing that I might shock it with more pickle juice. Or maybe I benefit from a blood quinine level derived from occasional tonic drinks diluted with gin.)

Another type of "cramp" is equally mysterious—exercise-related transient abdominal pain (ETAP). You likely know of it as a stitch in the side. When acute, it is a sharp, stabbing pain usually under the ribs on just one side of the abdomen and commonly encountered during torso-straight activities such as running, horseback riding, basketball, dance aerobics, and swimming. Episodes are more likely when exercising strenuously immediately after a meal or drinking a large volume of liquid. (To be clear, *recreational* swimming soon after eating is unlikely to cause either ETAP or muscle cramping.) ETAP more likely affects young athletes of either sex and is not associated with body type or level of conditioning. Here again, the cause remains unknown. Plausible theories include cramping of the diaphragm muscles, which are exercising strenuously to produce rapid breathing, or a jouncing of the intra-abdominal ligaments that support the stomach, spleen, pancreas, kidneys, and liver. Prevention strategies include not consuming a large volume of food or water for several hours before exercise, improving core strength, and wearing a wide, supportive abdominal band. Treatment recommendations include slowing down, pushing on the painful area, and breathing deeply. Since symptoms are transient and ETAP is self-limiting, its exact nature and best treatment have not been scrutinized.

What about fatigue? Surely there is better evidence and consensus for the cause of this common malady. Well, sort of. Experts divide fatigue into two parts, central and peripheral. Central fatigue is what happens in the brain—the uncomfortable feeling of being tired. But it occurs long before the muscles are actually exhausted. Incentives can overcome the psychological roadblock of central fatigue. The excitement of a major competition, the roar of the home-team crowd, and a trainer's urging all help summon capacity to go the extra mile.

One experiment had subjects "wall sit" for as long as they could. This meant backs against the wall, legs bent, and rear ends level with the knees to simulate a sitting position. When offered a paltry sum of money, the wall sitters quit after two minutes. When the reward was markedly increased, the average time doubled. Training can help get past this psychological barrier, and one athlete learned to say to himself, "Shut up, legs." It may also be that some individuals are just more pain tolerant than others, which would make them more resistant to central, psychological fatigue.

At some point, however, peripheral fatigue, which has nothing to do with psychology, sets in. For intense, brief activities such as sprinting or weight lifting, 30 minutes of rest often suffices to allow for muscle recovery. During this time the nerve endings at the neuromuscular junction replenish the chemical transmitters, and the muscles reestablish proper levels of sodium, potassium, and calcium. For endurance activities such as distance running, recovery from fatigue takes at least two to three days to restore glycogen deposits in the muscles and liver, for the nerves to recover their ability to produce sustained high-frequency signals, and for metabolic waste products to dissipate.

Statin medications can produce a spectrum of problems for skeletal muscles, from nonspecific aches of no consequence to full-blown, life-threatening autoimmune reactions. These medications reduce the liver's production of cholesterol and are therefore commonly prescribed to combat high levels of circulating cholesterol and the formation of plaques in coronary arteries. This treatment is often life-extending, but roughly 5 percent of people on statins experience muscle symptoms. Sometimes a cause-and-effect relationship is unclear, because people with or without statins on board get muscle aches from the flu and from exercise. Is the statin responsible, or is the juxtaposition merely coincidental? The prescribing doctor needs to find out immediately and withdraw the medication if a blood test shows unfavorable muscle changes. Genetic variation may make certain individuals more susceptible to statin myopathy.

Genetic Errors

The serious muscular defect known as muscular dystrophy (from Greek: *dys*: lacking, *trophy*: nourishment) has a genetic basis. It is either heritable or results from a new mutation. There are numerous forms of muscular dystrophy, differentiated by where a molecular disconnect occurs. Normally, the actin/myosin units in a muscle attach to a long protein that leads to the cell surface. Other proteins outside the cell continue the linkage so that the actin/myosin units ultimately connect to the tendon and farther on to the bone. The result is, voilà, muscle contraction and joint movement. The linkage is loosely analogous to that between a car's engine and the wheels. Whether a disconnect is because the transmission is in neutral or because the drive shaft is broken, revving the motor does not move the car. It's worse in muscular dystrophy, however, because any disconnection causes the actin/myosin units to tug on the cell membrane and tear it. Bad stuff leaks in and good stuff leaks out. The muscle cells die.

The most common form of this condition is Duchenne muscular dystrophy, named after the French neurologist who wrote about it extensively in the nineteenth century. Researchers far more recently have identified the disconnection to be the first long protein that normally links the actin/myosin units to the cell membrane. It is an X-linked, recessive disorder, meaning that it affects boys far more often than girls, who, if their other X chromosome is normal, are carriers for the condition but do not have the disease.

Duchenne muscular dystrophy appears about the time the child begins walking. It seems paradoxical that as a boy so afflicted grows older and progressively weaker, his calves appear more muscular than wasted. With the molecular linkage disconnected, muscle fibers gradually become disordered and die and are replaced by fibrous tissue and fat. That, along with the fact that the remaining muscle fibers try to take up the slack by hypertrophying, accounts for a paradoxically robust appearance. As weakness progresses, braces, crutches, and wheelchairs become useful. Duchenne muscular dystrophy also affects cardiac and smooth muscle. Late in the disease, heart failure and underactivity in the gastrointestinal and urinary systems occur.

Treatment with testosterone-like steroids, with their potent anti-inflammatory and muscle-enhancing properties, prolongs strength and the ability to walk for several years and extends longevity. (Life

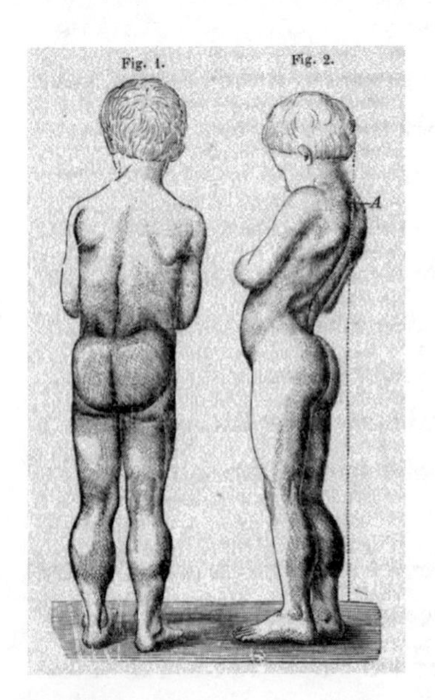

These images are from the original 1868 article describing the most common form of muscular dystrophy. The drawings exhibit the characteristic pseudohypertrophy of the lower limb muscles and the marked curvature of the lower back caused by weakened trunk musculature.

expectancy is as little as fifteen years.) Unfortunately, the side effects of treatment with conventional steroids are profound and include excessive weight gain, stunted growth, behavioral changes, and decreased bone density, which increases the risk of fractures. A synthetic steroid presently undergoing late-stage FDA-approved investigation appears to slow the progression of weakness in Duchenne muscular dystrophy and to have fewer side effects. Gene therapy is also on the horizon.

Another genetic muscle malady, malignant hyperthermia, can run in families but yet go entirely undetected for generations. Then it can emerge in an operating room and quickly lead to death. This is particularly shocking and newsworthy when an athlete, in peak condition, undergoes a "routine" procedure for an injury and dies during surgery. Certain commonly used anesthetic gases and relaxing agents given to susceptible individuals cause uncontrolled release of calcium from muscle cells, resulting in sustained muscle contractions. A downward spiral begins. This includes rapid depletion of ATP stores along with excessive consumption of oxygen and glucose. Exaggerated rises in carbon dioxide levels and body heat ensue along with rapid pulse and breathing. Multisystem failure leads to death in 80 percent of untreated patients.

Fortunately, the FDA approved a medication in 1979, and since then it is mandatory for it to be immediately available in all operating rooms where general anesthesia is used. The medication was discovered in the 1960s, and doctors used it initially to treat spasticity. Its value in treating malignant hyperthermia quickly became clear after researchers learned that the medication prevented catastrophic outcomes in genetically susceptible pigs. For humans, prompt administration of this drug and total body cooling with ice have reduced the incidence of malignant hyperthermia's mortality to less than 5 percent.

Of course, prevention remains preferable to treatment, but there are no symptoms or physical findings to suggest a person is susceptible. A muscle biopsy (obtained under local anesthesia) can reveal the genetic condition, but it is not done routinely. If you have had any family members suddenly die in the operating room, be sure to warn your anesthetist before going under. When there is the slightest suspicion that a patient might have one of the more than 25 genetic mutations causing malignant hyperthermia, alternate anesthetic agents are available.

Another alarming genetic muscle malady, this one causing slow death, is fortunately exceedingly rare, and I mention it to further highlight the spectrum of muscle diseases and to garner awareness for how skeletal muscle is, in general, remarkably resistant to serious disorders. In the instance of fibrodysplasia ossificans progressiva (FOP), muscle turns into bone. The trigger might be an injection, a bruise, or a fall. When surgeons first removed these ossified areas of muscle, they were aghast to see that more bone promptly formed in response to the operative injury. Skeletal muscle, tendons, and ligaments are all affected, and over years the individual becomes locked in bone. Finally unable to breathe because of limited chest expansion, affected individuals die of respiratory insufficiency.

In FOP, muscle is the victim, not the villain. The culprits are a protein that stimulates bone growth during fetal development and

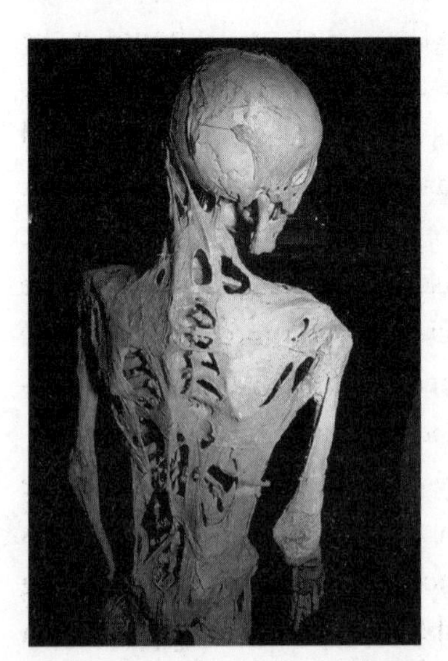

In this fortunately rare condition, fibrodysplasia ossificans progressiva, minor trauma causes bone to form in muscles, which progressively and irreversibly stiffens the musculoskeletal system and leads to early death from the inability to breathe.

another protein that usually blocks the first protein from generating new bone after the skeleton forms. A mutated gene erroneously codes the blocking protein and thereby allows the bone-forming one to run rampant in nonskeletal locations. There is a new bone-blocking drug that prevents extraskeletal bone formation in mice. Further studies, including trials in individuals with FOP, are underway.

More Molecular Maladies

One more mysterious condition, far more common than FOP, apparently not genetic, and fortunately not lethal, is neurogenic heterotopic ossification, which is medicalese for bone that forms in muscles, particularly around the hips and knees, following traumatic brain and spinal cord injuries. This poorly understood condition occurs when a neurologically injured patient is bedridden and perhaps on a ventilator. Presently the only treatment is surgical removal of the ossified muscles six months to a year after injury, when the patient's condition has stabilized. Scientists are studying various inflammatory mediators and stem cell activators to try to get a handle on this condition and find a way to prevent it with medications.

To understand the next condition, look back at the figure at the beginning of chapter 3 and note the dotted line labeled *fascia*. It is the thin, transparent layer of fibrous tissue that surrounds skeletal muscles and helps them hold their shape, as casings do for sausages. If there is a local injury, a fracture, for instance, within many of the body's fascial compartments, the fascia expands and makes space for leaking blood. Therefore, the increasing volume does not cause any pressure on the muscles contained within the fascial compartment. Particularly in the forearm and lower leg, however, the fascia is tough and resists any tendency to stretch. So rather than the compartment's volume increasing in response to leaking blood, its pressure does. (Consider the difference in overeating while wearing pajama bottoms with an elastic waistband versus wearing pants with

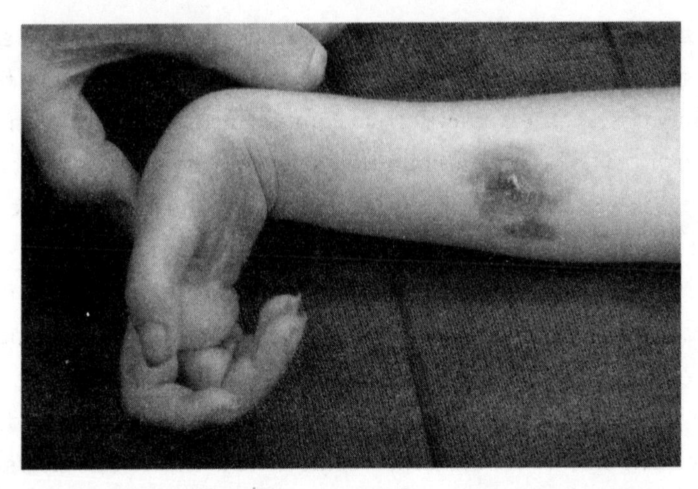

This boy sustained fractures in both forearm bones. Swelling ensued and was restricted by the fascia surrounding the adjacent muscles and by pressure from a tight cast (discolored area on forearm). The loss of circulation to the muscles caused their death and eventual replacement with unyielding scar tissue, which caused locking of the fingers in this dysfunctional position.

an unyielding tight belt.) *Compartment syndrome* ensues when the pressure increases to the point that it overpowers the arterial pressure and cuts off circulation to the enclosed muscles. If the pressure in the compartment is not promptly relieved, the muscle dies. Then, over months, dense scar tissue gradually replaces the dead muscle and prevents the joint(s) crossed by the muscle from moving. To relieve the pressure before muscle damage occurs, a surgeon must slit the fascia open, which allows for unfettered swelling, akin to cutting off a glutton's belt. In 1881, German physician Richard von Volkmann described the characteristic fisted posture of the fingers of such a patient. The contracture continues to bear his name.

A much milder form of compartment syndrome is chronic exertional compartment syndrome, which accounts for some cases of "shin splints" and "arm pump." In both conditions, pain and weakness develop with exertion and resolve with rest. Runners note these findings in their lower leg muscles. Motorcycle racers complain of pain in their forearm muscles shortly after they begin forcefully gripping the handlebars. The fascia prevents the affected muscles from swelling in response to exertion, so compartmental pressure increases and interferes with circulation to the oxygen-craving muscles. If rest, massage, and nonsteroidal anti-inflammatory medications are not curative, surgically opening the fascia often is, but not always.

The final molecular/cellular muscle disorder I want to describe is rhabdomyolysis. In Greek, *rhabdo-* means rodlike, *myo-* means muscle, and *lysis* means breakdown—that is, destruction of striated muscle, specifically the skeletal rather than the cardiac type. The condition most commonly occurs after a crush or compression injury that occludes the circulation to a limb for several hours. Doctors and emergency workers noted this in survivors of an earthquake in 1908 in Sicily, in German soldiers during the First World War, and in survivors of the blitz on London in the Second World War. All the victims had one or more limbs pinned down by heavy beams or broken slabs that cut off circulation. Known peacetime causes include earthquake entrapments, extreme exertion, overheating (malignant hyperthermia, for example), drugs (including antipsychotic medications and, rarely, statins), and prolonged immobilization while a comatose person is positioned on a hard surface.

Without any blood flowing to the involved area, the muscle cells die from lack of oxygen and break open (lyse). The cells' molecular contents, which are necessary and perfectly safe when contained within muscle cells, spill out. When the crush victim is rescued or a comatose person is moved, circulation to the area returns and carries the muscle's toxic remnants throughout the body. These toxins wreak havoc, especially on the kidneys, causing them to fail. Intravenous hydration and, in extreme cases, dialysis are effective treatments if rhabdomyolysis is recognized in time.

Rhabdomyolysis was described 15,000 years ago in the Old Testament book of Numbers 11:31–33. The writer describes a severe plague that descended on the Israelites during their exodus from Egypt and caused many sudden deaths shortly after the wanderers had consumed quail in bountiful quantities. Their symptoms were compatible with rhabdomyolysis. Various lines of reasoning conclude that this event occurred in the spring, when Mediterranean quail are known to feast on hemlock seeds. Historians and toxicologists have connected the dots because modern cases of rhabdomyolysis may occur following ingestion of Mediterranean quail in the spring. Why ingested hemlock is not toxic to quail yet causes rhabdomyolysis in humans who eat Mediterranean quail in the spring remains unknown.

All the skeletal muscle mishaps discussed so far occur on the molecular and cellular levels. I find it ironic that the precise causes for the commonly encountered conditions—growing pains, morning stiffness, delayed-onset muscle soreness, cramps, fatigue, and shin splints—are incompletely understood, whereas the exact gene mutations causing many of the rare and often lethal conditions have been fully worked out. Is it perhaps because the commonly encountered conditions are not threatening of life or limb and are therefore of less interest to researchers? Or is it that the causes are too subtle to be probed successfully with currently available methods? I find it exciting that muscle has secrets yet to be revealed. For now, I happily turn to *mechanical* muscle maladies, happily because the causes are obvious and the treatments often curative.

Mechanical Mishaps

The first is disconnection of a muscle from the skeleton so that its contraction no longer powers joint motion. The same happens when a bicycle's chain breaks. Pedal as hard as you like, the bike doesn't move. Tendon lacerations in the hand and forearm are obvious examples of musculoskeletal disconnections. Less commonly, a forceful stretch can tear a tendon or pull it off one of its bony attachments. It is quite rare that the muscle itself tears. Instead, the weak link where the disconnection occurs is the tendon or its bony attachment.

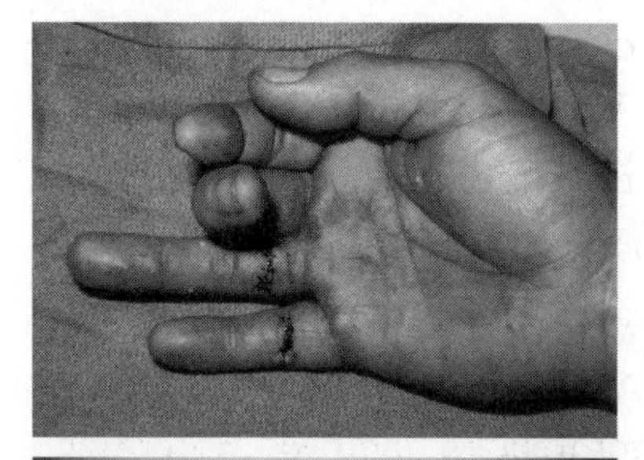

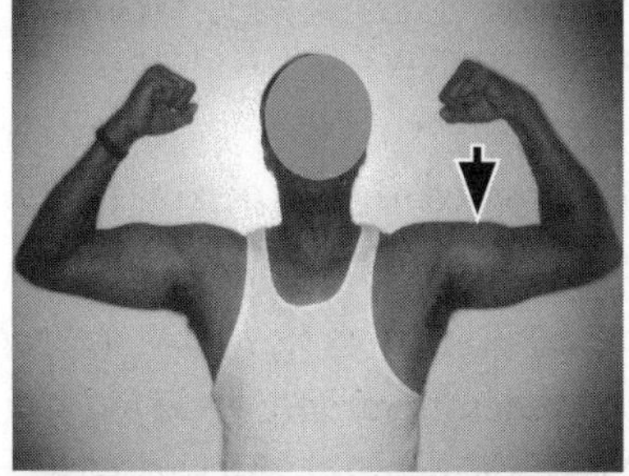

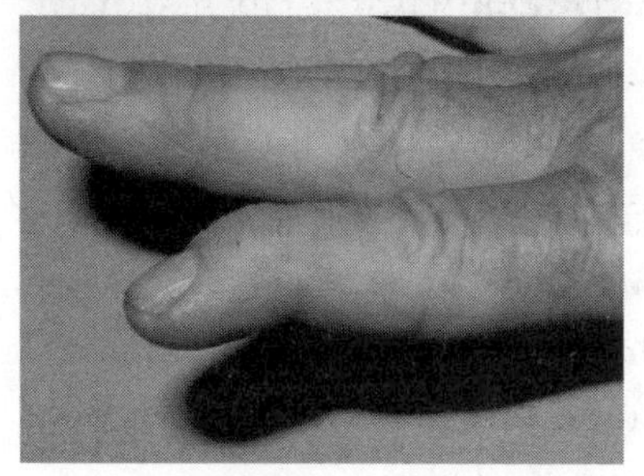

Examples of muscle/tendon disruptions.
Top: A knife injury cut the tendons that normally close the ring and small fingers.
Middle: Trying to lift a heavy object, this man pulled his left biceps off its bony attachment at the elbow. The muscle's bulge is still present, but it is now closer to his shoulder, and elbow flexion is weak.
Bottom: A volleyball injury stretched the tendon that normally straightens the joint nearest the fingertip.

A few tendon disruptions can heal with immobilization alone—for example, the tendon near the fingernail's base that straightens the digit's terminal joint. Many, however, need surgical reattachment, and following a period of rehabilitation, nearly normal function often returns. A torn heel cord (Achilles tendon) and a rupture of the biceps attachment near the elbow are two common examples.

In other instances, a muscle is in continuity but too long or too short to perform its usual function. The causes are multiple, sometimes genetic, sometimes resulting from injury or infection. Consider the six muscles that surround the eyeball and that are responsible for its movement in the eye socket. When the muscles are imbalanced, the eyes don't properly align when looking at an object.

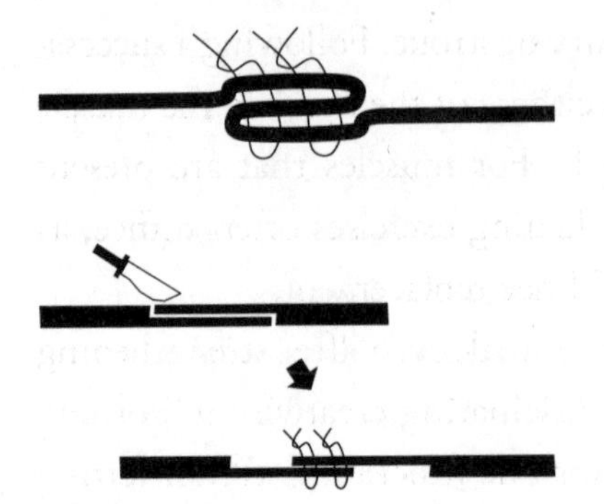

A muscle/tendon unit that is too long can be shortened by overlapping and suturing its tendon. One that is too short can be lengthened by dividing the tendon with a step cut and then reconnecting it.

If the offending muscle is too long, ophthalmologists can either detach, advance, and reattach it, or just take up the slack with several carefully placed sutures.

Sometimes a muscle is too short, which prohibits the joint it controls from opening completely. This is common in cerebral palsy, which might cause children to walk on their toes because their heel cords are too short to let their feet rest flat on the floor. Likewise, a too-short muscle can result from a stroke if the brain erroneously sends messages to keep a muscle contracted. Left untreated, the muscle gradually accepts its shortened condition as normal, and the capsule of the joint controlled by the muscle gets tight. Together they prevent the joint from moving through its normal range of motion. A common scenario is a stroke victim who cannot straighten elbow, wrist, and fingers. This makes it markedly difficult to slip the limb through a sleeve and causes the hand to rest in a clutched, awkward position in front of the breastbone. In such instances, the muscle-tendon unit can be rebalanced by making a "step cut" in the tendon, lengthening it, and suturing the tendon ends back together. Preoperatively, temporarily paralyzing the muscle with a botulinum injection often indicates what the permanent tendon-lengthening procedure would offer.

The same holds true in instances in which the muscle will stretch and contract sufficiently to allow for full joint motion, but it is just too strong and opposes the normal muscles that would ordinarily move the joint in the opposite direction. This can occur with cerebral palsy or after a severe head injury or stroke. Following a successful botulinum trial, a surgeon can either cut the nerve to the muscle or lengthen its tendon or detach it. For muscles that are present and functioning but weak, strengthening exercises often suffice, as might be the case following a total knee replacement.

For muscles that are absent or too weak even after strengthening exercises, surgeons have exhibited fascinating creativity in borrowing a muscle that normally performs one function and transferring its origin, insertion, or both, so that it substitutes for a more important and missing motion. These transfers reached a zenith of application following the polio epidemic in the 1950s when orthopedic surgeons were searching the body nearly head to toe to find muscles to transfer that could restore patients' ability to hold their unbraced knees straight or allow them to bring their hands to their mouths.

The surgeons' success was variable according to which muscles were paralyzed and which muscles remained strong and were relatively expendable to consider for transfer. For instance, muscles on the front of the upper arm (biceps and brachialis) cause the elbow to close and bring the hand into range of the mouth. The muscle on the back of the upper arm (triceps) straightens the elbow, but in a pinch gravity will also do this. Therefore, in the event of paralyzed biceps and brachialis muscles and the presence of a fully functioning triceps, surgeons would detach the triceps from its insertion at the back of the elbow and reattach it to the biceps tendon on the front of the elbow. Then the patient would learn to close the elbow by contracting the triceps. To straighten the elbow, the patient would relax the triceps and let gravity take over.

After polio was brought under control, tendon transfers were principally in the domain of hand surgeons who used (and use) them to restore function to badly injured limbs. Such a procedure, for example, can restore critical motion to the thumb by detaching a tendon from one of the other digits, rerouting it, and attaching it to the thumb. Patients quickly adapt to such changes and begin advantageously using the transferred muscle in its new position

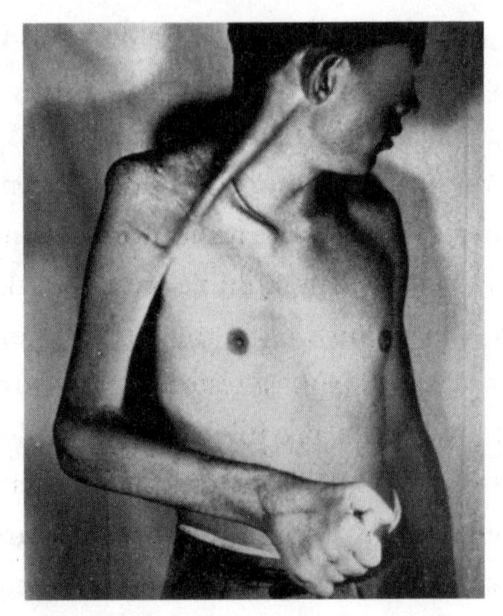

This image demonstrates a muscle transfer from the neck to the arm after the normal muscles for closing the elbow had been paralyzed from polio, and none remained around the shoulder that were strong enough to transfer. The strap-like muscle normally running from the breastbone and collarbone to the skull just behind the ear (sternocleidomastoid) has been rerouted and extended with a tendon graft, which has been attached to the biceps tendon. Contracting the muscle while holding the head still then closes the elbow. Turning the head away further tightens the transfer and increases elbow flexion.

without any conscious thought. An analogy would be moving a table lamp from a bedside table to a dresser without unplugging it. The lamp provides light to a previously dark area.

The continued development of microsurgery in recent decades has allowed surgeons of various specialties to perform "free" muscle transfers for a wide variety of infirmities. These procedures are by no means inexpensive. Rather, "free" refers to the fact that the surgeon completely detaches (frees) the donor muscle from its original position and divides its arterial, venous, and nerve supplies. Then the surgeon attaches the muscle to a new origin and insertion elsewhere in the body and reconnects the vessels and nerve to like structures nearby. This is akin to unplugging a table lamp in the bedroom, moving it to the kitchen, and plugging it back in.

A commonly used donor muscle is the gracilis, a long, slender muscle on the inner thigh. Once the gracilis is removed, the other muscles in the area can substitute for its normal function. The gracilis proves to be a workhorse in multiple areas—in the face, to substitute for paralyzed facial muscles; in the forearm, for restoring finger closure; and in the upper arm, to restore elbow closure.

A transferred muscle can also fill in, not for a missing motion, but for a large area of missing skin or for an abnormal contour, which might result from trauma or follow removal of a large tumor. In the case of missing skin, the muscle in its new location provides durable padding and is either covered with a skin graft or is transferred along with its overlying skin. As with muscles transferred to restore func-

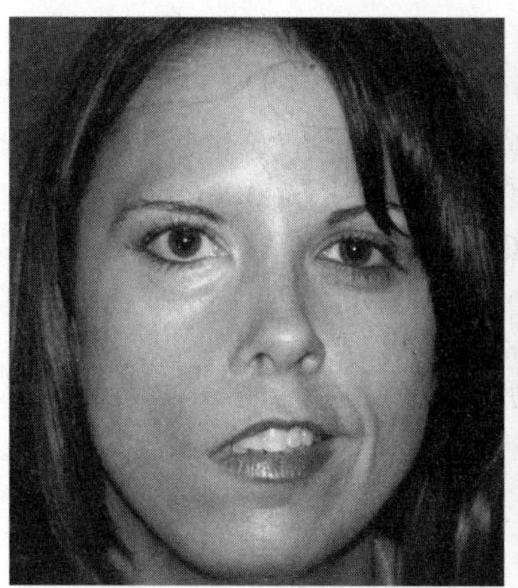

This woman had congenital paralysis of her right facial nerve, which caused her mouth to droop. The reconstruction included rerouting the functioning temporalis muscle, which is one of several muscles involved in chewing. The surgeon lifted the muscle's origin from the temple area on the side of the skull, folded it down and forward, and reattached it around the borders of the upper and lower lips. Initially after surgery, patients with this transfer have to bite their teeth together to contract the muscle and smile. With time, they unconsciously learn to contract the temporalis independently of the muscles that remain in place for chewing.

tion, muscles moved for coverage or contour can be left attached at one end and rotated locally, or they can be "free" transfers—detached completely and reattached in the needed location.

An example of a workhorse muscle frequently rotated for coverage is the "gastroc flap." Two calf muscles (gastrocnemius and soleus) perform the same function of raising the heel when standing and walking. When coverage is needed after a severe injury around the knee, half of the superficial calf muscle, the gastrocnemius, is expendable because the remaining half and the underlying soleus muscle remain in place and preserve the owner's ability to stand on their toes. The surgeon detaches the "gastroc" near the ankle and folds it up to cover an exposed shin bone, an open knee joint, or both.

In a similar fashion, for creation of the breast's contour following mastectomy, one-half of the six-pack muscle on the abdominal wall (rectus abdominis) often serves as donor tissue. The surgeon detaches it from its connection on the pelvis and, along with its overlying fat and skin, folds it upward onto the anterior chest. An alternative muscle flap used for breast reconstruction is the gracilis muscle. It, along with adjacent fat and skin, is cut completely free from the inner thigh and transferred to the breast area, where its blood vessels are reconnected.

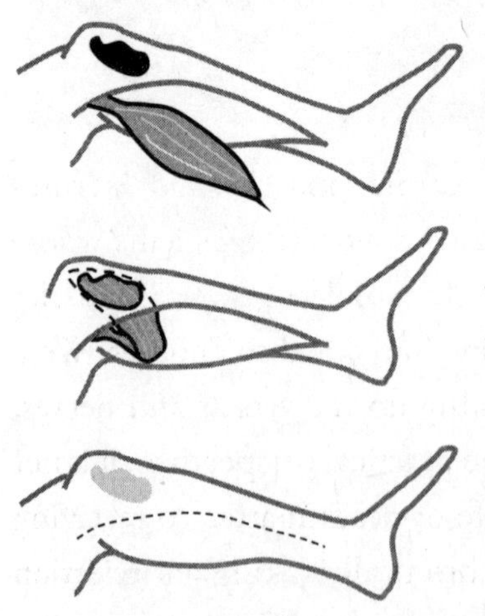

A large area of missing skin with exposed bone and an open knee joint (black) can be reconstructed using a gastrocnemius muscle flap. The muscle is detached from near the ankle, folded back on itself, and placed under the skin surrounding the defect. The exposed muscle at the recipient site is covered with a skin graft, which would not "take" over the open joint and exposed bone.

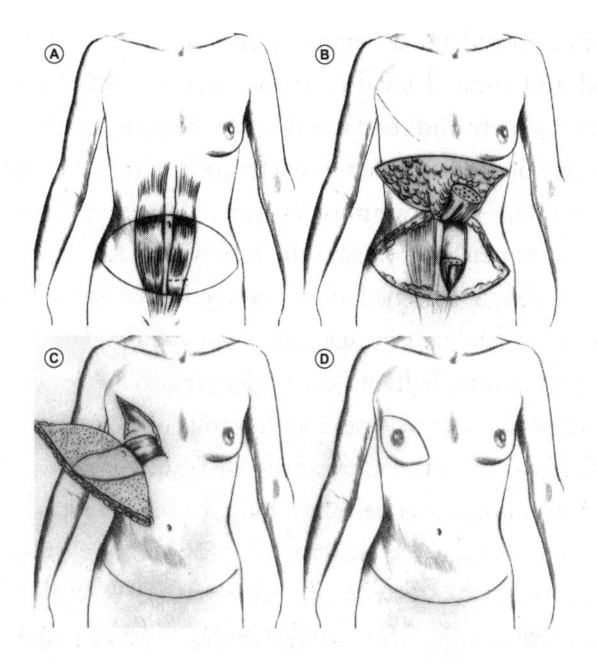

A. For breast reconstruction following mastectomy, a TRAM (transverse rectus abdominis muscle) flap is often used.
B. Detached from the pelvis, half of the abdomen's "six-pack" along with its overlying skin and subcutaneous tissue is turned and folded upward.
C. The flap is tunneled beneath the skin and used to restore the breast's contour.
D. The skin edges at the donor site are drawn together and sutured to leave a transverse "bikini line" scar.

Since surgeons transplant hearts (cardiac muscle) and uteruses (smooth muscle) from recently deceased donors to recipients whose own versions are inadequate, one might wonder if skeletal muscles can be similarly transplanted from one individual to another. Well, from a technical perspective of hooking up the vessels and nerves, they can; but from immunological and practical perspectives, it is not normally done. A failing heart is a life-or-death matter, so receiving a transplant even though it means chronically taking antirejection medications, which have their own set of risks and side effects, is a reasonable undertaking. Having a functional uterus is not a life-or-death matter. However, for those who feel that carrying one's own fetus is a transformative experience, the several years of required immunosuppression following a uterus transplant may seem justified. In addition, humans do not have any standby uteruses of their own that can substitute if the primary one is inadequate.

That's not true for skeletal muscles. As I have described above, the loss of function from a missing one may not be debilitating. If it is, there are many potential substitutes from the patient's own body, which avoids all immunological issues of performing an isolated human-to-human skeletal muscle transfer. The one exception is when an entire hand is missing, and a transplanted hand brings along its muscles with all the other tissues. True, a missing hand is not a life-or-death issue, but like the malfunctioning uterus, its absence reduces the amputee's quality of life, and artificial hands to date are lacking in some combination of appearance, dexterity, sensory awareness, and durability.

Surgeons worldwide have performed well over 100 hand transplants since the first one was done in 1998. Along with skin, bones, nerves, and vessels come the muscles in the palm, and perhaps the forearm, depending on the level of the original amputation. As nerves regrow into the transplanted limb over many months, some muscle function recovers, but never entirely. So yes, under special circumstances, skeletal muscle can be transplanted, but the lengthy surgery and the need for lifelong immunosuppression and extensive hand therapy make hand transplants extremely expensive.

Likely you had no idea of the varied problems that can affect muscle. Fortunately, the most serious ones are rather uncommon, and treatments for them seem to be under nearly continuous improvement. In general, muscle works flawlessly, and my aim has been to raise your awareness of how it silently serves us in so many ways. To ensure that our lovefest for muscle is comprehensive, we should sample some of the unique and stupefying ways it serves other animals.

Chapter 9

ZOOLOGICAL SURVEY

PUT YOUR LIPS TOGETHER AND HUM A BIT. NOW PINCH YOUR nose closed and continue. "Impossible," you say. Well, it is for humans but not for dolphins, porpoises, and toothed whales, which include the sperm and killer ones. They "sing" underwater without exhaling by use of their phonic lips, which are flexible muscle tissues in one nasal passage. Rather than exhaling after they have sounded off, they recirculate the air down the other nasal passage and cycle it repeatedly. It is only the nasal passage that is not endowed with phonic lips that terminates at the blowhole. Muscles surrounding the blowhole have self-evident Latin names, including *dilator nares* [nose] *superficialis* and *constrictor nares.* Farther inside, a cylindrically shaped muscle pistons in the nasal passage to provide secure closure of the airway regardless of the depth of a dive.

Hum again. To do so, you exhale. Try humming as you inhale. If there is any sound at all, is it more of a monotonal moan? That is not the case for some songbirds. They can sing whether they are inhaling or exhaling and can go on nonstop for hours. Some of these songsters can even produce two musical notes simultaneously. This comes about by a unique anatomical arrangement. Where the windpipe and the primary bronchial tubes meet, the airways become membranous and quite flexible. They can be independently opened and closed to varying degrees by the surrounding muscles, which allows separate notes from each bronchus to burst forth.

Elephants also make a wide variety of sounds, some of which

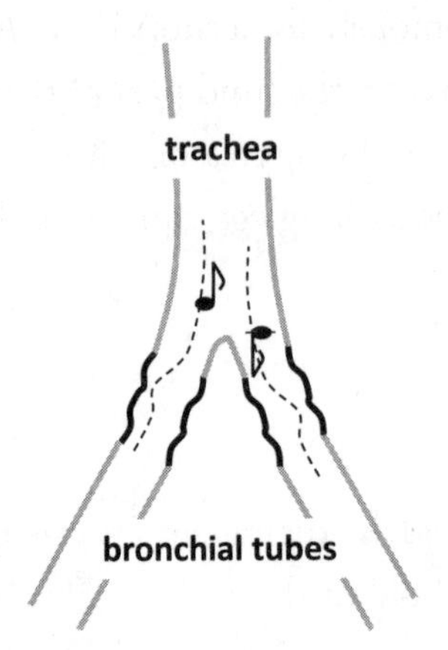

Some songbirds can differentially vibrate tympanic membranes located where the bronchial tubes and trachea meet and can thereby produce two notes at once.

come from their vocal cords, others from low-frequency rumbling exhalations through their trunks. Their trunks stand out in the animal kingdom, however, not for the sounds they can produce, but rather for their unique arrangement of muscles. They make this appendage the blue-ribbon winner for complexity, utility, and strength. This long, boneless, highly flexible nose has two centralized tubular airways extending its entire length. Surrounding the airways are four layers of muscle, two obliquely oriented, and one each lengthwise and circumferential. Together the muscles number 40,000 and account for the trunk's nearly universal flexibility and the elephant's ability to pick up a 700-pound log or a single peanut.

Equally impressive on a smaller scale are octopus and squid tentacles, each with multiple layers of muscles, which like the elephant's trunk are unimpeded in their motions by any bone. Furthermore, each limb has hundreds of independently controlled suckers that not only grasp but also smell and taste. Each sucker has muscle fibers aligned like lines of latitude and longitude on a globe as well as a third layer radiating from the center out, in starburst fashion. The complexity and functional capabilities are astounding. Although an octopus has a large brain for its body size, roughly two-thirds of tentacle control comes from nerve centers in the tentacles themselves. If an octopus loses an arm to a predator, the sacrificial limb and suckers will continue to function autonomously for a short time. As an escape tactic, some slugs can self-amputate the hind part of their foot, which wiggles violently and distracts the aggressor while the remainder of the slug slithers away. A daddy longlegs can similarly offer up a leg.

Tails

Remarkably, there are animals with widely divergent ancestries whose tails snap off when a predator has hold of them. The tail wiggles and provides the assailant a distracting snack while the main course escapes. Certain lizards (reptiles), salamanders (amphibians), and scorpions (arthropods) thereby survive. For scorpions, the separation eventually proves to be lethal, but not for several months, giving them an opportunity to reproduce. Salamanders so equipped survive and regrow an entirely normal tail fully equipped with muscle and bone.

Lizards generate a new but somewhat shorter tail over a matter of months, complete with muscles, although a cartilaginous tube replaces the bony vertebrae. How well the new appendage works is under scrutiny. Numerous zoologists have studied the sprint speed, stride length, stability during jumping, and climbing ability of lizards with original and regenerated tails and have come to conflicting conclusions. The differences may stem from which lizard species the researchers investigated and the exact nature of the obstacle course that they used. As with many areas of burning scientific inquiry, one paper on regenerated lizard tails concludes, "Further studies are required." (Job security!)

If you are feeling deficient without a tail, detachable or not, it gets worse. Cats and dogs curl and wave their tail muscles to communicate and indicate their mood. A cheetah uses its long tail for counterbalance and as an air rudder to assist with quick turns during a chase. Monitor lizards, crocodiles, and alligators thrash theirs around as weapons. Animals as diverse as seahorses, anteaters, opossums, porcupines, pangolins, and some monkeys and mice can curl their tails sufficiently to hang onto branches.

Tongues

A giraffe can also grasp, not with its tail but with its 18- to 20-inch-long tongue. This herbivore grabs acacia branches and pulls its head back to break off its next bite, thorns and all.

A woodpecker's tongue muscles originate on a nasal bone between its eyes. These muscles pass up and over the bird's forehead, around the back of its skull, and finally enter the throat. Hence, it is a very long muscle. A hungry woodpecker can extend its tongue to half its body length and retrieve succulent morsels from afar.

It is a family of salamanders, however, that earns my top respect for excellence in lingual performance. Its tongue has the highest acceleration and power output of any vertebrate movement. The extension measures 80 percent of the owner's body length, and it is equally efficient at body temperatures ranging from 36 to 75 degrees Fahrenheit.

If your tongue is feeling a bit inferior at this point, have it talk to your platysma, which is another muscle that is relatively insignificant in humans compared to its extent and function in most other mammals. In these animals it is the panniculus carnosus, literally

A woodpecker can extend its tongue beyond the tip of its beak about half the length of its body. On retraction, the tongue retreats into a channel at the back of the throat that extends over the top of its head and ends at the base of the beak.

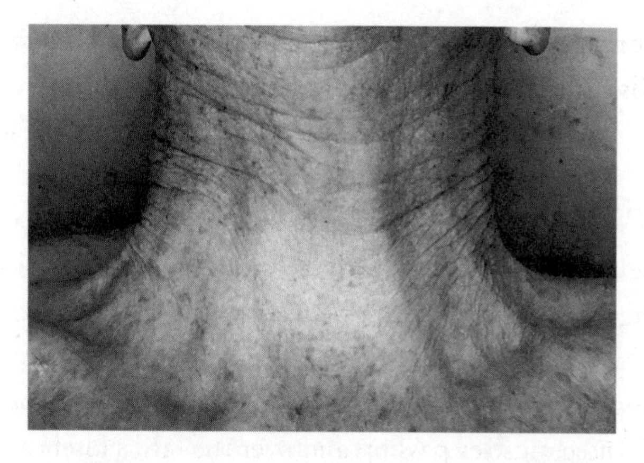

Contraction of the platysma muscle, which is attached to the jawbone and collarbones, tightens the skin on the anterior neck.

fleshy kerchief. It is a thin layer of muscle located immediately under and attached to the skin of the trunk and to variable degrees down the legs. It is easiest to see it at work on a horse. Should a fly begin to bite, the victim contracts its PC, which causes the skin to flicker, and the fly moves on. A nursing whale, unbothered by insects, can contract her PC to aid expression of milk from her mammary glands.

In humans the vestigial PC is the platysma (from Greek, *platusma*, flat piece, plate). It is present only on the front of the neck, running between the lower edge of the jawbone to the collarbones. Since we have other ways to deter biting insects and it is not positioned to express milk, our platysma doesn't get much of a workout, so it tends to sag with advancing age.

Various observers have lumped similar bone-to-skin muscles under the general term panniculus carnosus. Among them was Charles Darwin in his 1871 work, *The Descent of Man*:

> Rudiments of various muscles have been observed in many parts of the human body; and not a few muscles, which are regularly present in some of the lower animals can occasionally be detected in man in a greatly reduced condition. Every one must have noticed the power which many animals, especially horses, pos-

sess of moving or twitching their skin; this is effected by the panniculus carnosus. Remnants of this muscle in an efficient state are found in various parts of our bodies; for instance, the muscle on the forehead, by which the eyebrows are raised. . . . Some few persons have the power of contracting the superficial muscles on their scalps; and these muscles are in a variable and partly rudimentary condition. M. A. de Candolle has communicated to me a curious instance of the long-continued persistence or inheritance of this power, as well as of its unusual development. He knows a family in which one member, the present head of the family, could, when a youth, pitch several heavy books from his head by the movement of the scalp alone, and he won wagers by performing this feat. His father, uncle, grandfather, and his three children possess the same power to the same unusual degree. This family became divided eight generations ago into two branches; so that the head of the above-mentioned branch is cousin in the seventh degree to the head of the other branch. This distant cousin resides in another part of France; and on being asked whether he possessed the same faculty, immediately exhibited his power. This case offers a good illustration how persistent may be the transmission of an absolutely useless faculty.

Dogs have two well-developed muscles around their eyes that shape that portion of their face and allow them to be expressive. Wolves, by comparison, lack these muscles and therefore can only stare dispassionately.

One reason dogs are man's best friend is their expressive eyes. Dogs have two well-developed muscles that allow them to variably shape that portion of their faces. Wolves, by contrast, lack these muscles and seem to have entirely dispassionate stares. These two muscles are also absent in Siberian huskies, which genetically are more wolfish than poochish. Humans lack them as well.

Going Limbless

The specialized skeletal muscle functions described so far are all in animals with legs, fins, or wings. How do multicellular animals without appendages manage something so basic to life as movement? Snakes manage just fine. They most commonly undulate their bodies alternately from side to side and use fixed objects on land or the resistance of the water when swimming for push-offs. On sand, where there are no fixed objects, a proven strategy is to sidewind. The snake presses the ends of its body to the ground and lifts the intervening section to the side. Then, with its middle section planted, the sidewinder lifts its head and tail and advances them. Snakes have several other movement-effective strategies as well, but you get the idea, and at least they have bony skeletons against which to lever their skeletal muscles.

Scallops use their large muscle, the one we find so delectable when seared, to rapidly close their exoskeleton shell and squirt out water as a jet propellant. With separate strategies, slugs and earthworms move forward using their skeletal muscles, although here it is better to call their muscles *voluntary* rather than *skeletal* because these animals do not have skeletons.

The bottom side of a slug, called a foot, is well muscled and can contract in rhythmic waves to provide forward motion. Should you be curious to see this means of locomotion, place a snail or slug on a piece of clear glass or plastic and watch it from below.

Earthworms, also boneless, locomote by an entirely different means, the equivalent of which is already familiar to you—peristalsis, which is presently churning your most recent meal. A lowly worm uses its two sets of muscle, one looped around its body and the other one longitudinally arranged. When the worm contracts its circular muscles, it gets longer. Contracting its lengthwise-oriented muscles makes it shorter. To move, the earthworm anchors the front part of its body to the surrounding soil with tiny bristles and contracts its longitudinal muscles to pull the rest of the body forward. Then engaging the bristles on the hind part of its body against the soil, it squeezes with its circular layer of muscles to elongate its body and push its head forward.

Further appreciation of muscle's capabilities across the animal kingdom comes from noting some superlatives. For humans, I have already described a few athletic records that measure extraordinary performances, ones that principally require muscle-versus-gravity or muscle-versus-time endeavors. These are the kinds I like best because they do not rely on any technological advances such as improvements in vaulting poles, running shoes, or bicycles. Also, I prefer recognizing accomplishments that do not require judges with scorecards—too subjective and maybe even politicized.

Peak Performances

For other animals, I like their outstanding muscular performances even better because, for the most part, they result from a pure quest for survival—no adrenaline rush from roaring crowds, incentives from product endorsements and gold medals, or pistols going off behind their backs. Here is a sampling of some amazing ones that demonstrate the diversification and specialization of muscles seen in the animal kingdom.

Huaso, a horse, failed at racing, dressage, and show jumping before he demonstrated his prowess in high jumping. In 1949, at age 16, he jumped 8'1" to set a record yet unequaled. Consider too that Huaso performed this feat with his trainer on board. Wild animals,

The quarter-inch-long froghopper, a.k.a. spittlebug, is a phenomenal leaper, capable of jumping over two feet high.

however, likely sniff at this record because, although riderless, a cougar has high-jumped almost 20 feet and a dolphin almost 23 feet. Not so high but nonetheless impressive, a 50-ton humpback whale can propel itself completely out of the water at 20 miles per hour, which may be the single most energetic burst of muscular power in all of nature. You go, fast-twitch fibers!

Considering body length to height jumped, the froghopper (spittlebug), is the clear winner. It is a fourth of an inch long and can jump over 27 inches high, 100 times its body length. The froghopper could not achieve its powerful launch by sudden direct muscle contractions; rather, it uses a catch mechanism also employed by fleas and grasshoppers. The trick is that these critters slowly store energy and latch their hind legs in the "ready" position, in the same way a crossbow is slowly tensioned, restraining the energy until the instant of release. Trap-jaw ants do the same with their mandibles. Loading them in the "ready" position deforms the skull, and when the ant suddenly releases all the stored energy, the jaws snap shut with a speed comparable to that of a bullet—among the fastest movements any animal is capable of making. The trap-jaw ant uses its jaw "latch-mediated spring actuation" mechanism not only offensively but also defensively. Should a situation get dicey, the ant puts its head down, snaps its jaws shut against a solid surface, and launches itself tail first to safety.

For speed records, let's consider three categories—land, water, and air.

The extremely poisonous black mamba snake can move through its lightly wooded sub-Saharan habitat at 14 miles per hour. (If the mamba could keep that up for 26 miles, it would have the fastest marathon time ever.) That speed is particularly impressive since the mamba does it without legs. Legs matter, however, because an ostrich can sprint short distances at 60 miles per hour; by cutting its speed in half, it can continue for 30 minutes, thus demonstrating possession of an awesome combination of fast- and slow-twitch muscle fibers. If two legs are good, four are better. Cheetahs hold the land speed record at 70 miles per hour.

Leatherback turtles and gentoo penguins are the fastest swimmers among reptiles and birds, respectively, and can speed along at 22 miles per hour. The fastest mammalian swimmers, orca whales, manage 10 miles per hour. These muscular feats, however, are no match for fish. Bluefin tunas, mako sharks, and bonitos all zoom through water at about 40 miles per hour, and black marlin can double that speed.

In the air, horseflies clock in at 90 mph. For flying mammals, the record goes to the Mexican free-tailed bat at 100 mph. The appropriately named needletail swift outpaces the others at 106 mph. (Peregrine falcons dive at 200 mph, which must be a thrill; but that is gravity-assisted and therefore not in the running for muscle gold medals.)

When compared to the speeds I have just mentioned, 32 miles per hour seems leisurely, but what if a critter could fly at that speed nonstop for 7,000 miles over nine days? That is what the bar-tailed godwit does twice a year as it migrates back and forth from New Zealand to Alaska. Before embarking on its trans-Pacific flight, this 15-inch-long shorebird in the sandpiper family gorges itself, markedly increasing its fat stores but also bulking up its flight muscles.

Remarkably, the muscle enhancement occurs without exercise. Then, as takeoff looms, the godwit's digestive tract atrophies, which is acceptable since the bird neither eats nor drinks during its journey. How all this happens is a mystery, but when someone cracks the code, just imagine—humans bulking up without the need for exercise. In the meantime, the godwit's migration deserves recognition as the supreme muscle-endurance performance in the animal kingdom.

Bite force is another category of superlatives among animals. The sizes of the body, jaw, and jaw-closing muscles along with the muscles' locations along the jaws' lever arms all contribute to the bite force. Even tooth configuration matters because pointed teeth will exert their force over smaller areas and therefore measure greater pressure than flat teeth. Quantifying bite force is riskier than measuring the height of a froghopper's jump, so imagine what it took to obtain the following results: Humans chomp down with an average force of 162 pounds per square inch. Hyenas do so with 7 times more force, bull sharks with 8 times, hippos with 11 times, and Nile crocs with 31 times more force. Other strong biters include jaguars, gorillas, bears, and Tasmanian devils.

Horsepower

In the eighteenth and nineteenth centuries, a horse's strength became an important norm for comparison. It was the age of steam, and inventors, engineers, industrialists, and entrepreneurs needed a standard measurement of power against which to rate machines in a way that had conceptual relevance to potential buyers. Scotsman James Watt was instrumental in describing an engine's output in "horsepower." Watt determined the amount of weight a harnessed horse pulling a rope, strung over a high pulley and attached to a pallet of bricks, could lift in a given amount of time. He equated one horsepower with the ability to raise 550 pounds one foot in one second. This was the level of work that a horse could sustain over a day in the field and remain healthy. For such an endurance endeavor, horses would use their slow-twitch muscle fibers. For sudden power bursts needed for Kentucky Derby performances, horses call on their fast-twitch fibers, and their output approaches 15 horsepower.

Champion human weight lifters performing deadlifts, which entail lifting the barbell off the ground to the level of the thighs, do so in less than a second and exert 2 horsepower. Jamaican Usain Bolt produced a momentary burst of 3.5 horsepower during his world-record-setting 100-meter dash in 2009. Conditioned endurance athletes can manage about a tenth of Bolt's power output but can perform that level of "work" far longer than Bolt's burst.

I wonder if Watt envisioned that one day describing horsepower would mean carefully differentiating a mustang from a Mustang and a bronco from a Bronco?

✦✦✦✦

Over the centuries, artists and anatomists, often one and the same, extended their depictions of muscle anatomy to nonhuman subjects. Perhaps it was the power, size, and familiarity of horses that made them particularly enticing subjects. The melding of the art and anatomy of equestrian muscles is as apparent and beautiful as it is for human anatomy. Three artist-anatomists who portrayed well-muscled horses stand out.

Brit George Stubbs (1724–1806) was a brilliant portraitist. Horses were his forte. Stubbs dissected a dozen equines after suspending their dead bodies with slings and ropes and posing them in lifelike positions.

Parisian Antoine-Louis Barye (1795–1875) sketched living animals at the botanical garden's zoo and availed himself to the zookeepers when an animal died. He translated this knowledge into bronze sculptures, both large and small, of robust wild animals and mythological creatures often locked in mortal combat. An art

Late-eighteenth-century British artist George Stubbs specialized in painting horses, which he posed in lifelike postures. In preparation, he repeatedly dissected and drew equine muscle anatomy in its various layers.

critic of the time titled him Michelangelo of the Menagerie and wrote, "Barye aggrandizes his animal subjects, simplifying them, idealizing and stylizing them in a manner that is bold, energetic, and rugged."

Another Frenchman, Ernest Meissonier (1815–1891), particularly enjoyed painting and sculpting images of modern warfare and military life and was known for his authoritative draftsmanship and marked attention to detail. To achieve this with horses, he sketched from horseback every nuance of muscles in motion as he followed his models around the riding area that he had constructed especially for this purpose.

Nineteenth-century Frenchman Antoine-Louis Barye was renowned for sculpting well-muscled animals and converting them to bronze sculptures, both large and small.

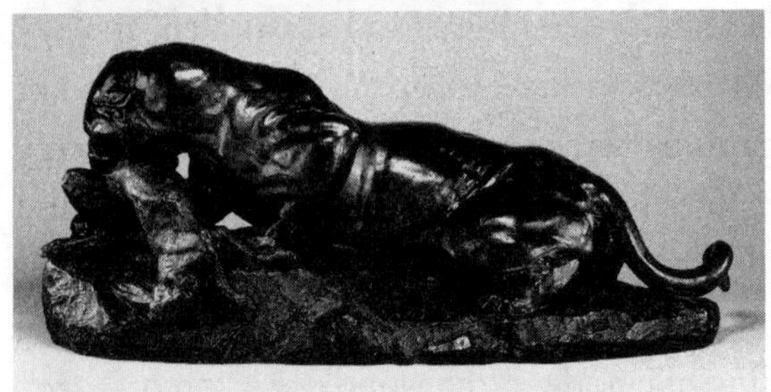

Chapter 10

OTHER FORCE PRODUCERS

UNTIL NOW, I HAVE FOCUSED ENTIRELY ON MUSCLE, WHICH IS the most obvious way that animals move. It is a force producer, a motor. Muscle's power emanates from the repeated interactions of actin and myosin molecules. I hope you agree that this dynamic duo, in their smooth, cardiac, and skeletal forms and across all of zoology for at least 600 million years, is hard to match. But muscle is not the only way that organisms convert thermal and chemical energy into movement. Other force producers in nature are even older, and in recent times humans have made artificial ones. To truly appreciate muscle, we need to understand a little about its force-producing competitors before titling one of them the world's best motor.

Probably from soon after their appearance 4.1 billion years ago, organisms have included actin, which raises a reasonable question. Between then and a mere 600 million years ago when actin first teamed up with myosin to make muscle, what roles were actin filaments fulfilling? The answer hinges on the definition of life, summarized by the now familiar MRS GREN, which, to remind you, stands for movement, reproduction, sensitivity, growth, respiration, excretion, and nutrition. For something to be considered living it must perform *all* these functions. Movement, for instance, occurs in tides and breezes, but these phenomena are under lunar and solar influences, respectively, and are obviously not alive. By comparison, actin, starting long before joining up with myosin, generates *purposeful* movement, *within* individual animal *and* plant cells, even if the cells themselves remain stationary. And for animals, both single and multicell varieties, actin accounts for purposeful movement as they navigate their environments. It's the same for amoebas in pond water and skin cells in panda bears.

Actin Acting Alone

Atomic forces cause each actin molecule to fold into a glob. The globs assemble themselves into a filament. When a cell wants to go somewhere—say a skin cell wants to creep across an abrasion to heal it—hundreds to tens of thousands of actin molecules line up

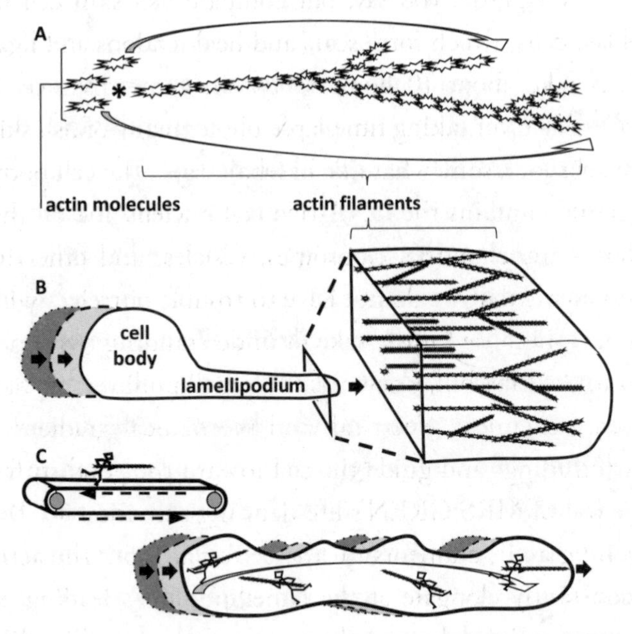

A. The chemical bonds between the actin molecules on a filament's trailing end weaken and break. This frees the actin molecules to move to the filament's leading end, reattach, and add branches.
B. An enlarged view of the lamellipodium shows actin filaments branching at their tips, which pushes the lamellipodium forward. The cell body follows.
C. This phenomenon is called "treadmilling" but is actually more akin to "tank treading" since a treadmill does not move forward but a tank does.

inside the cell to form molecular "sticks" to prod the cell membrane forward. The filaments can disassemble as easily and quickly as they assemble. Their life span is seconds to minutes.

Molecular biologists have learned much about animal cell mobility by studying skin cells from fish. Scientists like these cells because they tend to move in a straight line and so can be easily followed across a microscope slide compared to cells that tend to wander erratically. Also, the fish skin cells move quickly, which is obviously a benefit to fish when trying to heal cuts. But quickly is a relative term, because a fish skin cell takes 30 to 90 minutes to travel a sixteenth of an inch. "That's not very fast," you say, but compare fish skin cell motility to fibroblast cells, which form scars and heal tendons and ligaments. Fibroblasts take about 10 hours to go the same distance. Hence, microbiologists favor taking time-lapse photography of fish skin cells.

These cells look somewhat like baseball caps. The cell body is the cap itself and contains the DNA-rich cell nucleus and all the other organelles—mitochondria, ribosomes, vesicles, and other interesting components that we do not have to trouble ourselves with presently. That is because for the sake of understanding movement, the action is in the cap's bill, known as the lamellipodium, or "flat foot." It can sense chemical, electrical, and mechanical gradients in the cell's surroundings and guide the cell toward the good stuff to perform the rest of MRS GREN's life-defining activities.

Branching actin filaments fill the cell's flat foot. The actin molecules constantly elongate at the lamellipodium's leading edge by adding more actin globs and thereby prod the lamellipodium forward. The rest of the cell gets dragged along. Simultaneously, the filaments lose actin molecules on their trailing ends. Then these recently detached actin globs race forward and rejoin on the filament's tip. Molecular biologists call this phenomenon treadmilling, but I prefer to think of it as "tank treading." On a treadmill, the runner (cell body) stays in one place. That misses the point. Instead, the body of a tank (cell body) goes wherever its treads (lamellipodium) take it—toward the good stuff.

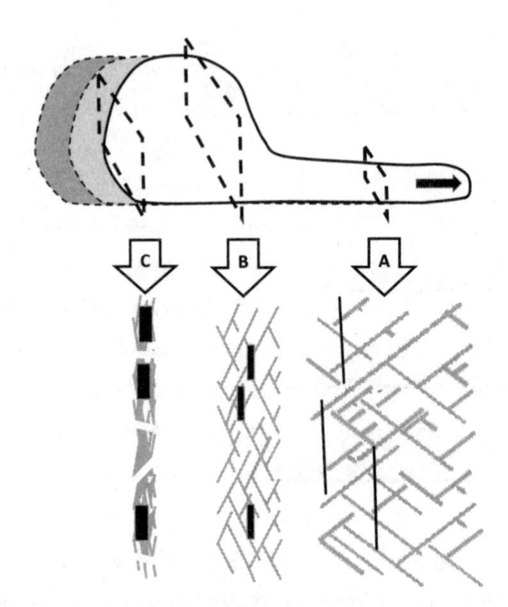

The cell is moving to the right, drawn by the advancing lamellipodium (black arrow). Cross-sections through three portions of the cell show the changing roles of actin and myosin.
A. Abundant branching actin filaments are pushing the lamellipodium forward. Thin myosin filaments (black) are sparse and inactive.
B. In the cell body, myosin filaments are contracting and altering the shape of the actin scaffolding.
C. At the cell's trailing edge, the fully contracted myosin breaks the actin scaffolding apart.

Actin filaments pack the flat foot, and hundreds to thousands of other proteins bind to the actin filaments and facilitate both their formation and destruction. One of the most important facilitators is actin's not-to-be-forgotten friend, myosin, which works in the zone where the lamellipodium and the cell body merge. There, myosin molecules grab onto the trailing ends of the actin filaments and tear them apart.

That this rapid assembly/disassembly of actin filaments happens in an organized manner and has been doing so for billions of years boggles the mind, even more so when one considers that a fish skin cell is so small that 20 of them could line up on a grain of table salt.

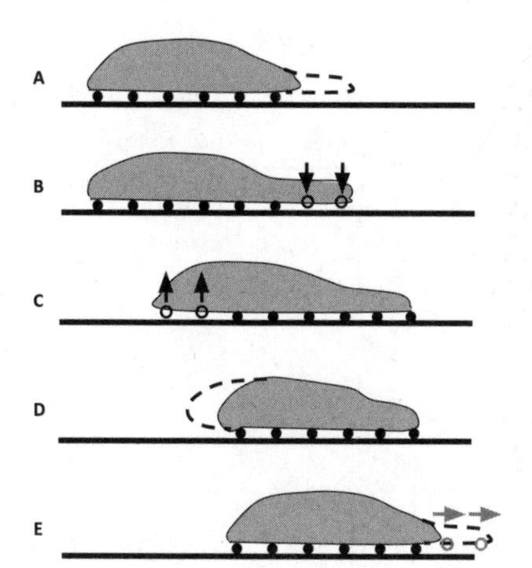

Cell migration by lamellipodium extension. A. The lamellipodium extends. B. New focal adhesions attach the "flat foot" to the surface. C. The adhesions on the trailing edge release. D, E. The cell moves forward and repeats the process.

That's a lot of molecular activity going on in a small space, which if nothing else begins to help us realize how tiny molecules are.

Like a car spinning its wheels on ice, a cell that is tank treading without traction doesn't move. It should come as no surprise that nature has this issue worked out as well. The leading edge of the lamellipodium attaches with multiple tiny weld spots to whatever it is advancing across—be it a healing wound or a microscope slide. The lamellipodium uses these attachments to pull the cell forward, and as the cell creeps along, the no-longer-needed attachments at the rear of the cell disappear. Consider the analogy of a tank slipping on ice. Foot soldiers throw buckets of sand in front of the spinning treads, and once the tank has gained traction and advanced, the foot soldiers come along behind and sweep up the sand. The details on how cells form and dismantle these focal adhesions, however, presently elude microbiologists. I find some satisfaction in this unknown because for as much as scientists have probed life's secrets with mind-boggling breadth and intensity, there are still unanswered secrets. Trivial? Absolutely not. Wouldn't it be great if there were a way to prevent cancer cells from gaining traction and using their actin filaments to move themselves about indiscriminately?

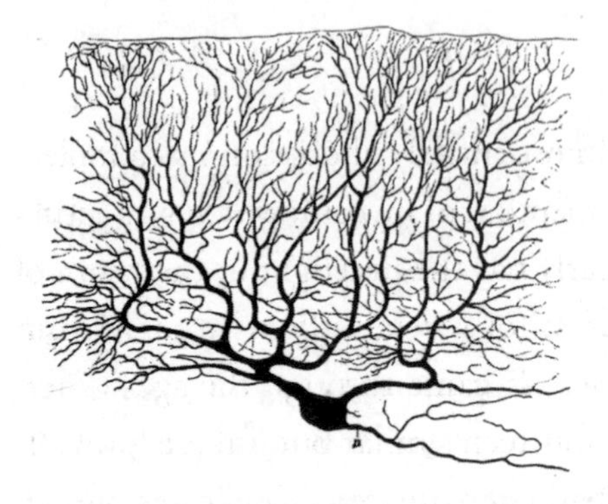

Nerve cells are particularly adept at sending out filopodia, which interconnect extensively with adjacent cells and make a dense neural network.

Regardless of whether a cell is benign or malignant, how does its lamellipodium sense the environs and know where to lay down its attachments and guide the cell's migration? It has "antennae," properly known as filopodia, which are thin, actin-rich tentacles that explore the surroundings and lay down adhesions to draw the lamellipodium and the cell body toward the good stuff. For instance, during development, a nerve cell's filopodia probe the environment, follow chemical attractants, and fan out (like electrostatically charged hairs on a cat) to form a dense network. They make myriad cell-to-cell contacts and contribute to the multiplicity of ways that electrical impulses can pass through the nervous system.

Actin-powered migrating cells routinely encounter environments far more forbidding than a thin layer of salty water on a glass slide. In real life, cells can dry out on the surface of a wound. Deeper down, they must wedge their way between cells that may not appreciate the intrusion. The filopodia and lamellipodia may be able to elbow their way in, but dragging the cell body and especially its rather rigid and unyielding cell nucleus behind them is a chore. Remarkably, ingenious experimentalists have measured the force produced by a single actin filament as it thrusts the cell membrane forward. Each filament pushes with roughly the same force that you would experience if you chopped an apple into a trillion pieces and then tested the heft of one bit on your fingertip. There is strength in numbers, of course, as the fact that cells do move around testifies—for instance, those that heal wounds and those that spread cancer.

Versatile Myosin

So far, this chapter has been almost entirely about actin as a molecular motor with only a passing mention of myosin, which was actin's coequal in chapter 2. Actin clearly has the limelight in the realm of cell motility. It has been steadfast for several billion years. Myosin has some interesting roles as well. It came through the ages differently and now exists in at least fourteen similar but unique forms. It is as if the players on one baseball team are exact duplicates of each other and can play any position competently (actin), while the other team has a specialist for every situation, including bunting, relief pitching, and pinch running (the myosin team).

The first form of myosin to be discovered and the one that is best understood is the one that steps along actin filaments in muscle to cause contraction. This form is *conventional* myosin. It is also integral to another process that we all have some observational experience with—wound healing. The critical cell here is the myofibroblast—a cross between an actin-and-myosin-laden muscle cell and the ubiquitous fibroblast, which is a collagen-producing wonder, responsible for interconnecting all our cells and preventing us from being nothing more than a puddle of cell soup.

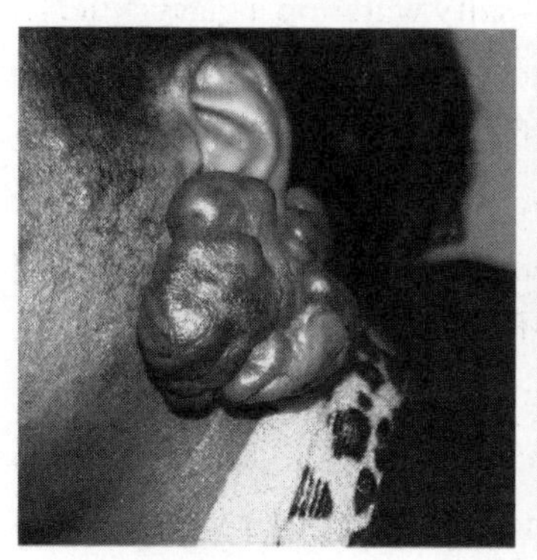

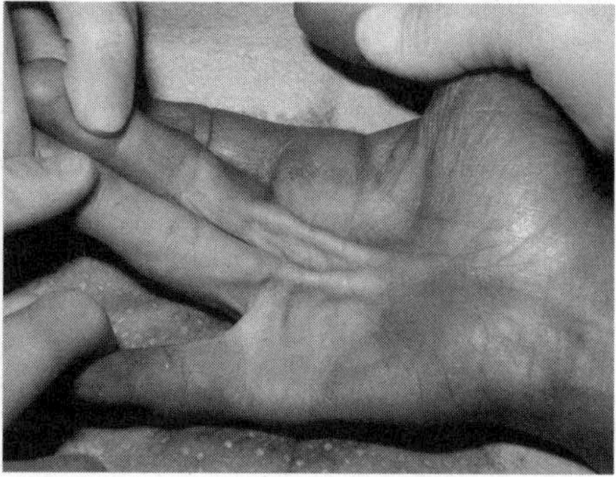

Left. This large keloid is the result of exuberant myofibroblast activity following an ear piercing.
Right. Myofibroblasts in Dupuytren's disease form cords that may progressively shorten, limiting flat-hand activities.

Myofibroblasts contract, as do muscle cells, but with only about a tenth of the force and without nervous system control. They populate open wounds, hold onto one another with strong intercellular connections, contract, and gradually draw the skin edges together. When the skin finally seals, the myofibroblasts disappear. An analogy may help. Imagine a scattering of canoes floating in a gigantic inflatable wading pool. If the canoeists closest to the pool's sides grab its edges while their partners paddle and they all pull hard, they can pull the pool edges closer together. As the pool gets progressively smaller, the canoes crowd one another, and so one by one they must sink or otherwise disappear for closure to continue.

Occasionally the myofibroblasts don't disappear after they have closed the wound. In such instances, the actin and myosin no longer contract, but collagen production continues unabated. This is the mechanism for keloid formation—wound healing that doesn't realize that it is past time to quit.

There are other conditions in which myofibroblasts misbehave and cause unwanted contraction, but in these instances it is not a wound but a genetic predisposition that incites the process. As

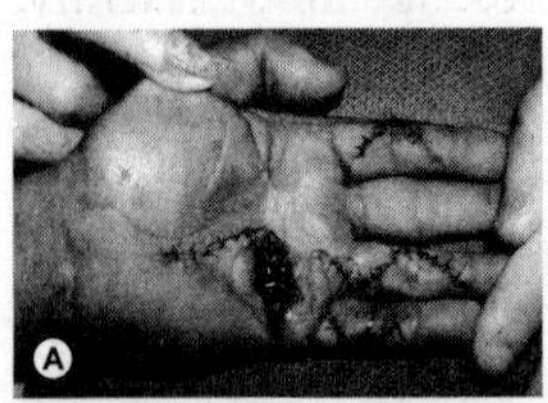

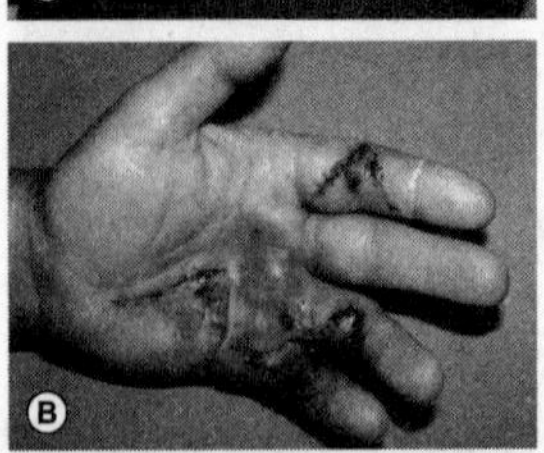

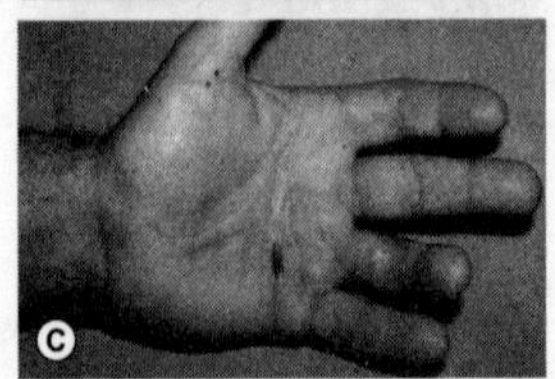

A. Following excision of tight Dupuytren's cords, the palm is left open to avoid stretching the skin acutely.
B. After 13 days, the myofibroblasts have slowly stretched the skin and reduced the width of the wound by about half.
C. At 35 days, myofibroblast contraction has nearly pulled the skin edges together.

they age, about 8 percent of men with northern European ancestry develop fibrous thickenings in their palms, which over many months to years turn into cords, tent the skin, and gradually shorten. The thickenings are benign, but they may become irksome or disabling if the tight cords prevent the fingers from straightening completely and thereby interfere with flat-hand activities such as clapping, face washing, and hand shaking.

The same genetic predisposition may also cause formation of fibrous thickenings over the knuckles or in the arches of the feet. These lumps may be disturbing but are not dysfunctional. What is dysfunctional, however, is when a thickening of this sort forms in the shaft of the penis, leading to what colloquially is known as "bent spike syndrome." These myofibroblast-induced thickenings in the hand, foot, and penis are called Dupuytren's, Lederhosen's, and Peyronie's disease, respectively. Many people have never heard of them, partly because the people who have these conditions may not be inclined to mention them.

The other functions that the myosins perform are not immediately evident to the naked eye and require special staining and labeling techniques in order to be appreciated microscopically. Presently, many of these functions are incompletely understood.

An exception is animal-cell division: A contractile ring of actin and myosin filaments forms around the middle of the parent cell.

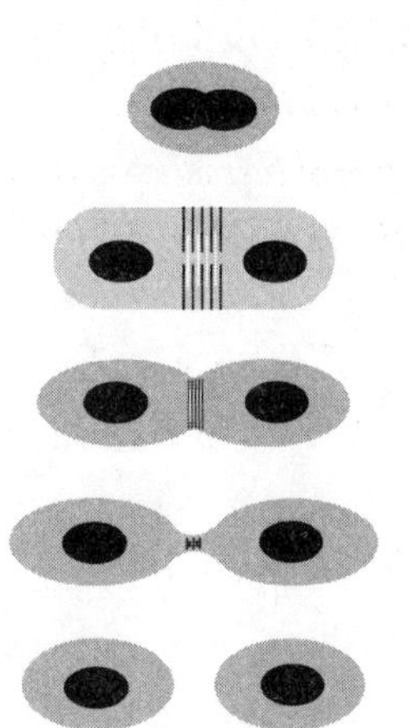

As cell division begins, a central ring of actin/myosin units encircles the cell. By progressively contracting, the ring gradually pinches the cell into two cells. In the process, the ring shrinks and ultimately disappears.

The ring progressively shrinks until it pinches the cell into two daughter cells. Remarkably, as the ring constricts, it does not get thicker; and when cell division is complete, it disappears entirely—another of molecular biology's marvelous feats.

All the conditions discussed so far—wound closure, the disorders of excessive fibrous tissue formation, and cell division—involve conventional myosin, the same form that occurs in muscle. Many of the unconventional forms work as single filaments and have stepping capability on one end, while their trailing end attaches to various subcellular organelles and molecules and transports them around the cell. Some of these unconventional myosins characteristically walk their cargo along actin filaments *away* from the cell nucleus and toward the cell's outer reaches. Others are "delivery trucks" bringing material from afar to the "inner city." Different categories of proteins regulate cargo loading and unloading, and it all works with speed and precision that would dazzle Amazon Prime.

For instance, an unconventional myosin in a tobacco plant travels at a speed that would carry it across a grain of salt in forty seconds, meaning that it could get clear across the cell that contains it in several seconds. Even more impressive is that this myosin walks along its actin "highway" at a rate of 200 steps a second, each cycle requiring the expenditure and reacquisition of an ATP energy packet. Not all myosins move that fast, but consider that a human consists of roughly 30 trillion cells and that myosins are dragging substances and structures around in every one of them. At any speed, I'm awestruck by the beauty and versatility of these molecular motors that are unobtrusively powering life.

The "S" in MRS GREN is for sensation, and the "N" is for nutrition. Actin and unconventional myosins have roles in both of these vital functions too. Projecting into the fluid in the inner ear are two arrays of tiny hairs. Sort of like seaweed attached to the ocean floor and moving with the currents, one array of hairs responds to the physical force from sound, and the other array reacts when the head changes position. Each hair is rather stiff due to the actin and myosin packed inside. So rather than swaying like seaweed, the hairs bend at their bases when mechanically perturbed, toggle-switch style, and generate electrical signals that pass either to the brain's auditory cortex or to the area that affects balance. The distance the hairs must move to stimulate an impulse is just a few nanometers (billionths of a meter), even for loud sounds. That's tiny. If you gave each of these micro bristles a bending radius of 5 nanometers, you could array 900 million of them on one surface of a grain of table salt, and they would never touch each other even when responding to a thunderclap.

Similarly sized actin-myosin-rich hairs project from the cells that line the small intestine and kidneys. They markedly increase these tubes' surface areas—in the gut to facilitate absorption of digested nutrients and in the kidney to aid reabsorption of water and sodium. You may remember from biology class that villi, fingerlike projections, line the small intestine. When observers first saw these through a microscope, they noted that each villus had a "brush border." Later, when scientists magnified the kidney and intestinal tissues many times over with electron microscopy, those minuscule paintbrush-like bristles were clearly seen to be "microvilli."

By the mid-1800s, scientists using enhanced light microscopy and 500X magnification could see movement *within* cells, which they described as "cytoplasmic streaming." This phenomenon convinced them that cells were the fundamental unit of life. The movement is particularly noticeable in large plant cells—nine of which would fit on a grain of salt. For these cells as well, unconventional myosin molecules step their way along actin filaments, which crisscross the cell. They transport various molecules and subcellular structures over distances too great for simple diffusion to be effective.

Two More Mini Motors

I would like to thank and shake hands with the sadly unknown person who came up with the memory device MRS GREN. For one thing, I find it useful for remembering living organisms' basic functions. But more important, the mnemonic honors muscle by placing the "M" for motion first because, in my mind, motion takes priority. Nonetheless, and despite actin's and myosin's agelessness, ubiquity, durability, and versatility, they are not the only molecular motors at work in cells. Two of these other force producers, kinesin (*ki NEE sin*) and dynein (*DIE neen*), are proteins that walk along intracellular "pathways" similar to the way myosin repeatedly advances and retreats along actin filaments. The pathways for kinesin and dynein are microtubules, which are arranged tightly inside every plant and animal cell and participate in cell division and help support cell structure. The microtubules also provide pathways for dynein and kinesin to transport various subcellular units and molecules around, which they do to keep each cell functioning properly. I think about dynein and kinesin as the low-slung tractors at the airport that tow airplanes and baggage carts around. In this analogy the taxiways, access roads, and aprons represent the microtubules, except that in the case of cells, the microtubule paths course in three dimensions.

Kinesin and dynein mirror each other's activity. Kinesin has two feet that walk along a microtubule pulling cargoes many times its own size *away* from the cell center and toward the cell membrane.

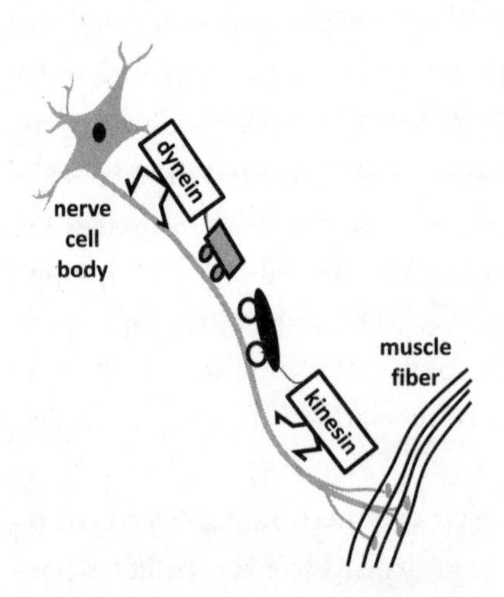

Dynein and kinesin are protein motors responsible for intracellular transport of subcellular elements and molecules. Dynein moves its cargo toward the cell center. Kinesin works in the opposite direction.

Dynein does the opposite, tugging cargoes from the hinterlands into the cell's center. Certain fish and amphibians use this give-and-take ability to quickly and repeatedly change color. Kinesin pulls the pigment-bearing organelles into small patches, which minimizes their influence on skin color. Dynein spreads the organelles out to maximize their effect.

Growing evidence suggests that mistakes kinesin makes, either by dropping its physiologic cargo or erroneously delivering it, may contribute to neurological diseases such as Alzheimer's. Kinesin can also mistakenly deliver nonphysiologic cargoes such as viruses deep into a cell. On the upside, doctors are considering using kinesin inhibitors to slow down or halt rapid cell division in cancers. The thought is that if intracellular transport of essential elements is retarded, then so is the cancer's spread.

Dynein is also the motor that powers movement in cells equipped with cilia and flagella. These are the hairlike appendages arising from a cell's surface that work like tiny paddles. They can move the cell through a moist environment or move something across the surface of a cell if the cell is fixed in one place. Although cilia are smaller and multiple and flagella are larger and sometimes singular, their microstructure and activity are the same. Had biologists known about the identical composition of cilia and flagella when they first identified them, one name would have sufficed, but we are stuck with two. Regardless of what we call them, it is dynein walking along microtubules that accounts for their action. In single-cell organisms such as paramecia and sperm, the appendages provide a way for the cell to get about in a manner that is far faster and more maneuverable than that of cells that are dragged along by their lamellipodia. Dynein-powered flagella not only speed sperm along and enable animal reproduction, they are also a fertilization feature for certain plants, including ferns and ginkgo trees.

As well as aiding motility in tiny creatures, cilia are present on stationary cells in multicellular organisms. Here they take respon-

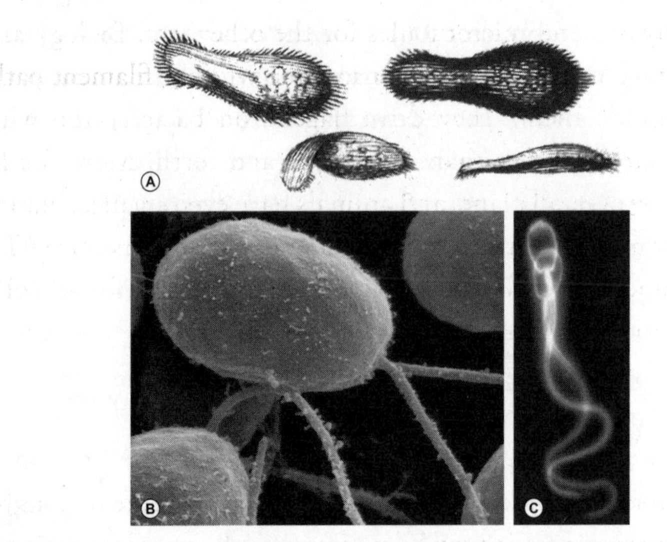

A. Slipper-shaped and single-celled, the freshwater animal Paramecium is covered with cilia, useful for locomotion.

B. Green algae sperm with single flagellum that spins to provide movement.

C. A time lapse of 0.28 seconds shows a ram sperm's flagellum powering it forward.

sibility for moving along whatever material is touching their surface. For example, the cells lining the trachea are ciliated and move in coordinated waves to transport dust and phlegm upward into the throat. There the material gets swallowed and disappears down the esophagus to the stomach, where gastric acid destroys it. We also have cilia on our nasal linings, and they serve the same purpose. Lining the cavities in the brain, ciliated cells keep nutrients in the cerebrospinal fluid circulating; and in the fallopian tube, the cilia hasten an unfertilized egg along toward its destiny. Logically explained then are dynein disorders, such as infertility from immobile sperm, ectopic pregnancies when a fertilized egg is not rhythmically waved into the uterus, and bronchial pneumonia caused by underperforming cilia in the lungs.

So far, all the nonmuscle molecular motors discussed—myosin, kinesin, and dynein—work linearly along filaments, actin in the case of myosin and microtubules for the other two. Biology also sports rotary motors, which manage fine without filament pathways to guide them. They drive flagella on bacteria and whip about in circles rather than wave back and forth. More fundamental, however, all plants and animals have even smaller rotating molecular motors that add a phosphate to ADP to reenergize ATP. This, I hope you recall, is the primary energy source for all MRS GREN functions.

Plant Movers

Plant motors are as fascinating as animal motors and are responsible for pushing roots through soil (sometimes heaving up sidewalks in the process), twining tendrils, opening flowers, flinging seeds, and even snatching insects. Plant motors, however, work in an entirely different way than animal motors do. They expand and push rather than contract and pull. This is possible and necessary because a cellulose-rich and rather rigid covering, the cell wall, encases each cell. Animal cells, by contrast, do not have cell walls, just their entirely flexible cell membranes. Plant cells can expand and shrink only a bit and do so by altering their water pressure. Plant cells can stand a far wider range of pressures without bursting than can animal cells, but they cannot deform and slip through small spaces as animal fibrous cells and cancer cells can do, pulled along by their lamellipodia. The rigid walls also prevent plant cells from entertaining any thought of contracting, an activity that muscle cells do so well.

For these reasons, any semblance of rapid motor activity of plants easily captures our attention. For instance, flowers turn to face the sun as it moves across the sky, leaflets close when touched, and leaves open and close according to the day/night cycle, an activity that intrigued Charles Darwin (1809–1882). He studied these actions in detail and confessed to a friend, "I think we have *proved* that the sleep of plants is to lessen the injury to the leaves from radiation. This has interested me much, and has cost us great labour, as it has been a problem since the time of Linnaeus [1707–1778]. But we have killed or badly injured a multitude of plants." In 1880, Darwin and his son, Francis, a botanist, compiled all their observations in a book titled *The Power and Movement of Plants*. Since then, scientists have learned much more about plant motors but still do not fully understand them.

What is known is that plant cells can absorb or release water by pumping sodium, potassium, and similar ions in and out. Water molecules are attracted to the ions and follow them. The cells use ATP as the energy source and generate energy with the same rotating molecular motor that humans use to pump iron. When a line of swelling cells is restrained on one side and at both ends, the line bends because it can't get longer.

Movement of this sort occurs over minutes or hours but cannot explain the lightning-fast movement seen in the carnivorous Venus flytrap and other predatory plants. In a tenth of a second, the flytrap's leaves snap shut on an unsuspecting insect. This is far faster than what any plant's ATP-powered ion pump can produce. Instead, the flytrap slowly loads the cells in a specialized pair of leaves with water, which bows them convexly. (You can get the idea by pinching the top and bottom edges of a playing card slightly toward each

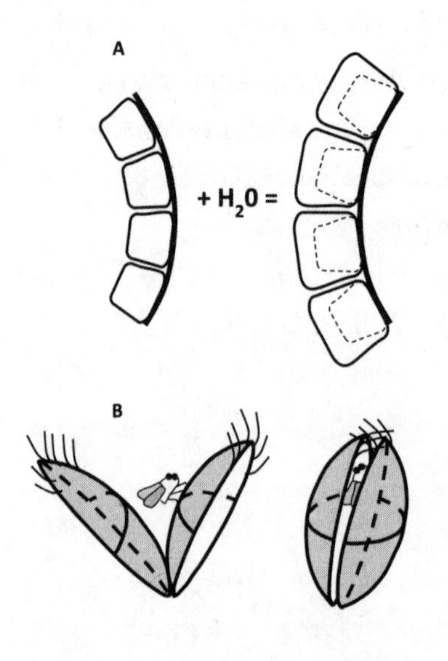

A. Plant cells swell with the addition of water, but when asymmetrically restrained, the cells will bow the stem or leaf.
B. The Venus flytrap's specialized leaves slowly take on water and become convex on their upper surfaces. When mechanically triggered by an insect, the leaves use the stored energy to snap suddenly from a convex to a concave configuration, thereby trapping the prey.

other to produce a bow. Then without reducing the pinch, you can take your other hand and flip the bow from convex to concave.) The Venus flytrap does the same. The water pressure has made the leaves elastically unstable and prepared them to snap instantly from their convex open position to their concave deathtrap.

These preloaded plant movements are all-or-none, whereas the speed and degree of animals' muscle contractions are variable. Nonetheless, physicists and engineers are interested in learning more about plant motors as inspirations for modeling robots. Here's why. Many plants spend their entire lives where their seeds germinated. Since they cannot move away to avoid environmental hostilities, plants are highly adapted to survive, and they do so in dual environments—air and soil, air and water, or water and soil. Because plants are successful under such hybrid conditions, they are good models for engineers to use when designing robots that must function in unstructured and unpredictable situations.

Stand-Ins

Presently, however, robots are powered by motors that act more like animal muscles than plant motors. Take, for example, exoskeletons,

A full-body exoskeleton allows the user to lift and carry superhuman loads without risk of muscle injury.

which are upper-body, lower-body, or whole-body "suits" that users strap to their bodies. The exoskeletons, powered by electric motors or perhaps pistons, provide the users with superhuman strength and endurance for activities such as repeated heavy lifting or prolonged overhead positioning. In the workplace, exoskeletons have the potential of increasing productivity and reducing the incidence of musculoskeletal injuries. Also, the US Department of Defense is experimenting with exoskeletons that would allow ground troops to carry packs weighing several hundred pounds. Exoskeletons additionally have the potential to help people with spinal cord injuries who are too weak to stand and walk on their own.

◆◆◆◆

PROSTHETISTS ARE APPLYING the same motorized hinged-limb technologies to the development of artificial limbs, particularly hands. These artificial parts replace missing fingers or whole hands rather than augmenting the function of intact but weak ones. Control of these robotic hands, however, is more complicated because the user needs to activate the prosthesis with contractions in existing muscles. Ideally, the messages to control the prosthesis would transmit directly from brain waves detected inside a special headband to the artificial limb. The best that today's technology can do, however, is to have the prosthesis read nerve impulses that the user sends to existing muscles, which the prosthesis senses and uses to activate a motor.

Until recently, this messaging was far from intuitive because, for example, an artificial hand might be set up so that when the user flexed the biceps, surface electrodes placed nearby would sense the electrical activity and tell the fist-making motor in the prosthesis to close the fingers. The required mental gymnastics are hard enough just to think about and even more difficult for an amputee to execute when performing a fine-motor skill, such as picking up an egg without breaking it.

Now surgeons are beginning to use an intriguing technique called targeted muscle reinnervation, TMR. They identify ends of nerves

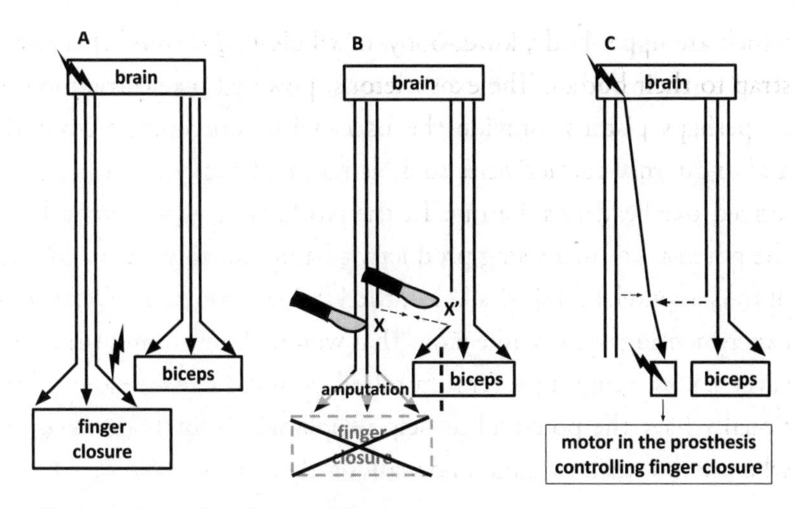

Targeted muscle reinnervation.
A. Ordinarily, the brain sends electrical impulses along separate nerves to individual skeletal muscles, which contract in response to the stimuli.
B. After an amputation through the forearm, the muscles controlling finger closure are absent, but the nerves that led to them remain functional. Here the surgeon cuts the nerve that previously controlled finger closure (X) and prepares to transfer it to a nerve that controls a portion of the intact and functioning biceps muscle (X').
C. When the nerve transfer has healed and the patient's brain says, "Make a fist," the message is routed down the original nerve to the small and isolated portion of the biceps muscle. The electrical impulse in the muscle, which is much stronger than the impulse in the nerve, is picked up by surface electrodes in the prosthesis and activates the motor in the prosthesis that closes the artificial fingers.

in the amputation stump, ones that prior to injury supplied now-missing muscles that performed a certain motion—for example, the muscles that would normally cause the hand to make a fist. The surgeons then reroute each nerve ending into a portion of an uninjured muscle in the upper arm or shoulder region. When all is healed, the patient merely has to think "close my fingers." The message goes forth on the original nerve and stimulates a portion of that nerve's new muscle to contract. That muscle's resultant electrical activity then triggers a motor in the prosthesis to close the artificial fingers.

This may sound complicated, but it is entirely intuitive to the user, who no longer must think, "When I want to close my artificial fingers, I tense my biceps. When I want to open my fingers, I'll raise my shoulder." With TMR, the practically unconscious effort that results in forming a fist is no greater with the prosthetic hand than it would be with the original. Think of TMR as turning the muscles in the amputation stump into a video game controller for a robotic hand.

◆◆◆◆

A LOT OF innovation is going into the design of artificial muscles that activate these power suits, robots, and prostheses. An ideal design would feature long excursion, high efficiency and durability, rapid operation, light weight, and low cost. Electric motors and air-actuated pistons have well-established track records, but in some instances high-tech alternatives to better meet the design criteria will replace them.

One is a shape-memory metal, also known as smart muscle, smart alloy, or muscle wire. Two metals, for instance, nickel and titanium, are combined. Bend the resulting alloy and it stays deformed. Heat it slightly, and it returns to its original shape. As an example, a shape-memory wire can be formed into a coiled spring and then stretched. Application of an electric current heats the wire slightly and returns the spring to its original length. It works repeatedly, but it is expensive and slow.

More promising alternatives are thermoactivated polymers rather than thermoactivated metals. Even nylon fishing line works. When combined with a fine wire and twisted repeatedly until it coils like a springy telephone cord, it shortens when an electric current passes through.

Just like real muscle, the longer the contrivance is, the more excursion it has; and the more coils that line up side by side, the stronger it is. Such a "muscle" is known as a "soft robotic" because it does not contain any rigid plastic or metal parts and is therefore more compatible when working adjacent to skin, but it is not ready to work inside the body like a real muscle, at least not yet. For that, tissue engineering has a growing role.

When somebody has lost a large segment of muscle, say from a war wound or tumor excision, the idea of an engineered replacement comes to mind. As an example, tissue engineering works well for cartilage damage on a weight-bearing surface of the knee joint. The patient donates some small plugs of cartilage, typically from the kneecap. The cells are "expanded" in the lab, turning thousands of cells into millions, which some weeks later are injected into the cartilage defect. The cells fill the "pothole," grow together, and restore a smooth gliding surface. The same process works well for major skin defects, particularly after burns.

For three reasons, engineering a functional muscle, however, is more problematic. Muscle's metabolic demands are far greater than either skin or cartilage, so if a sheet of engineered muscle is more than tissue-paper thick, it will need some engineered capillaries so it can receive nutrients and discharge metabolic waste products. Second, for an engineered muscle to be useful, it must contract and relax, so it must be flexible. Third, it will need a nerve supply to control its contractions. These seem like daunting challenges, but researchers are making progress on all three issues and have shown some promising early results in rats. Imagine the day when somebody wants to look ripped but doesn't want to pay their dues in the gym. They could have their muscle biopsied, send the cells to the lab where they are expanded millions of times over, and three months later have some new bulgy new muscles slipped in over their existing ones.

Another area of active research is the development of artificial muscles, which are already appearing in robots and exoskeletons. These units are usually electric motors or pistons driven by oil or air. (Spiders worked this out ages ago. They have only one set of muscles in their legs—ones that fold their knees and ankles closed. Spiders then straighten their legs by inflating them with blood, as if inflating a balloon.) The holy grail for artificial skeletal muscles, just like for artificial hearts, is fully implantable and biocompatible

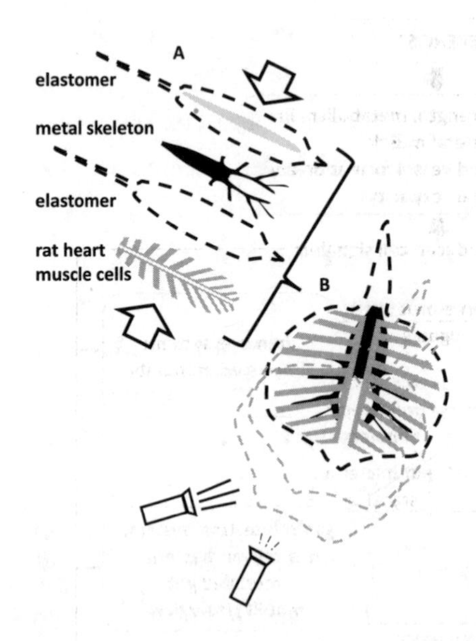

A. This cyborg stingray consists of two layers of silicone elastomer sandwiching a gold skeleton and supporting a network of rat cardiac cells.
B. The muscle cells respond to the energy imparted by two flashing lasers and ripple the wings to move the hybrid forward and also turn it.

ones, with an addition. The owners need to be able to voluntarily control their new skeletal muscles via nerves connecting them to the brain, while artificial hearts can just pump away without conscious consideration.

Finally, there are hybrids, which use living muscle cells to power fabricated components. Presently, these are just laboratory-based proofs of concept. If they ever become practical, they will have to function in a saltwater environment identical to muscle's natural home to keep the cells happy. The coolest one I have seen is a nickel-sized stingray cyborg that has thin silicone wings sandwiched onto a gold "skeleton" along with some rat heart cells. The muscle cells are encoded to respond to flashes of laser light, which can direct the "animal" to move forward or turn left and right.

Muscle's Messengers

Scientists and engineers are devising muscle substitutes to produce movement, but muscle is about much more than movement. Some researchers even describe muscle as an endocrine organ, which like

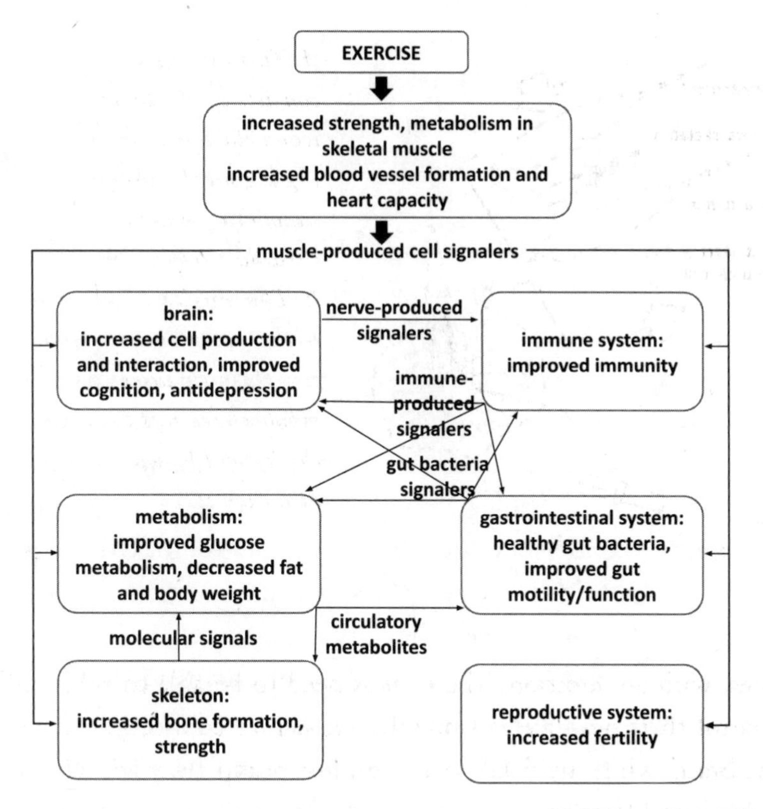

Exercise has well-known beneficial effects on skeletal muscle and on the cardiovascular and skeletal systems. Emerging research indicates that muscle-produced cell signalers also have wide-ranging and favorable effects on most other organ systems.

the thyroid, pituitary, adrenal, ovary, and testis, secretes chemical messengers that affect a variety of distant tissues, organs, and systems.

It is well known, for instance, that exercise induces bone health and cardiac health. I mentioned earlier that blood flow restriction therapy strengthens muscles in not only the treated limb but also in the opposite, nontreated limb. This observation implies that beneficial muscle-derived chemical messengers are circulating throughout the body. Experiments indicate that exercise ameliorates, if not prevents, a wide range of disorders including cancer, cardiovascular, cardiorespiratory, and metabolic diseases such as diabetes. Exercise also has beneficial effects on neurological conditions including Alzheimer's disease, Parkinsonism, stroke, multiple sclerosis. schizophrenia, depression, and dementia. Administering blood from old and regularly exercised mice into old and sedentary ones transfers the beneficial effects of physical activity on brain function to the inactive ones.

Once these chemical messengers are identified, couldn't we just pop a pill and receive all the benefits of exercise without having to sweat? Well, that will likely be a long time coming, because all our organ systems work in concert. Merely swallowing a "violin" without considering what the "trumpet" and "oboe" are contributing will not produce the desired harmonious symphonic masterpiece. To continue the music analogy, researchers are just beginning to understand the "score" and some of the contributing "notes"—the muscle-derived chemical messengers that enhance the health of other tissues. They call these muscle mimetics, and such a panacea pill would likely first be useful for individuals with mental or physical disabilities that prevent them from exercising. Most of us, however, need to keep on regularly and vigorously contracting our muscles, for their own sake as well as for our overall good health.

The World's Best Motor

As we come to the end, it is worth asking, "Just how good is muscle?" Could an artificial contrivance ever substitute for it, if not universally, at least broadly? Conversely, could muscle, as demonstrated by the stingray cyborg, be scaled up to power lawn mowers and cars? And do kinesin and dynein have future roles delivering parcels for FedEx? Unbiased as I am, I boldly claim that muscle is the world's best motor. My view is based on these criteria that I assembled: durability, scalability, ubiquity, versatility, adaptability, efficiency, practicality, and aesthetics. The world's best motor should rank high on all the scales. Let's look at these features one at a time.

How many electric motors or internal combustion engines remain

Hardly bigger than a grain of table salt, tardigrades, a.k.a. water bears and moss piglets, are remarkably resilient. Their muscles continue to function even after having been exposed to pressure extremes ranging from deep ocean trenches to outer space and temperatures ranging from hundreds of degrees cold to hundreds of degrees hot.

fully operational after 100 years of continuous service, or maybe even 500 years if you compare them to the longevity of Greenland sharks? Furthermore, contracting actin/myosin units have been around for millions of years and in tardigrades have proven to be incredibly resilient.

Tardigrades, also known as water bears or moss piglets, are tiny, weird eight-legged creatures. If one crawled onto a grain of table salt, it would slightly hang over the edges. Tardigrades may be the most durable animals of all. They survive pressures from the vacuum of outer space to six times greater than what is present in the deepest ocean trenches. They also endure withering radiation and temperatures ranging from –458 to +300 degrees Fahrenheit. Neither starvation nor dehydration vexes them. No motor other than actin/myosin comes close to enduring such extreme conditions. So for the criterion of durability, I give the nod to muscle.

For scalability, muscle serves a fairy wasp, which is even smaller than a tardigrade, and a blue whale, the largest animal ever. In the manufactured world, there are small electric motors and internal combustion engines that can rest comfortably on a fingertip, but a whole family of fairy wasps could rest on the tiniest artificial motor. At the other extreme, there are piston engines that power supersized container ships and which you could not hide behind a two-bedroom house. But neither could you hide a blue whale in a similar location. Therefore, for scalability, muscle again gets top billing.

Regarding ubiquity, muscle loses out to dynein, kinesin, and ATP synthase, which are integral to intracellular activity not only in all animals but also in plants, bacteria, and fungi. Muscle wins a second-place ribbon here because bar-headed geese use them to fly over the Himalayas 23,000 feet high and Mariana snailfish take them down 27,000 feet in the ocean.

For versatility and adaptability, I have spent the previous nine chapters extolling muscle's marvelous capabilities throughout the animal kingdom. Furthermore, muscle works on land, sea, and air and can power up on a variety of bizarre energy sources (corn dogs, haggis, raw oysters). By contrast, an alternating-current electric motor cannot run on direct current and an internal combustion engine cannot run on electricity.

Judging efficiency is nuanced. A motor that was 100 percent efficient would convert its entire power supply of electrical or chemical energy into motion. Electric motors come the closest, rating 90 percent efficiency in laboratory conditions and up to 60 percent in actual use, while the rest of the delivered energy dissipates as heat and friction. Internal combustion engines may be 50 percent efficient in the laboratory, but on the street they convert 17–21 percent of energy derived from gasoline into motion, with the rest going out as friction and as heat in the exhaust pipes and radiator. Automobile energy efficiency also depends on the driving pattern. Repeatedly flooring the accelerator and stomping on the brakes turns more of the energy into heat.

It is the same with muscle, which is 18–26 percent efficient in converting food into motion. Sudden fast-twitch-muscle-fiber activities such as skipping rope are less efficient than slow-twitch-muscle-fiber contractions such as walking. For mammals, at least, it is good that muscle is not entirely efficient, because the heat muscles generate keeps cellular activity throughout the body percolating steadily along.

Regarding practicality and aesthetics, the call is more personal. An engineer or cardiologist might choose a different winner than a prosthetist or botanist, but for me there is nothing more practical and enchanting than muscle.

Probably not to your complete surprise, I award the title of the world's best motor to—drum roll, please—muscle.

Cool Down and Stretch

Muscle is clearly a timely and captivating topic for investigation, description, and celebration, from molecular biology and nanoscience to what finally causes trained cyclists to fall from exercise bikes in exhaustion. I have great respect for the curious, committed, and intrepid scientists who through the ages performed anatomical dissections and recorded their findings. I also honor those who devised and completed ingenious experiments both in the laboratory and the clinic. We owe them all a debt of gratitude. Please take time to at least scan the reference citations in the Selected Bibliography. (These citations constitute half of the sources I referred to. The complete bibliography is available at MuscleandBone.info.) Look at the names of the international panoply of contributing authors and of the wide-ranging journals that have published their peer-reviewed research.

Despite this impressive body of existing research, there is still much to learn. For instance, what's behind the Kohnstamm phenomenon, demonstrated by your arms floating away from your trunk after you press your hands against a doorway? Can we come to fully understand fatigue and delayed-onset muscle soreness and then find ways to prevent these common maladies? How will explorers to Mars keep their muscles from turning to mush en route? What you have learned by reading *Muscle* should inform your understanding as new discoveries unfold.

Please don't forget MRS GREN. It is obvious that molecular motors are responsible for movement (M), but as we've seen throughout this book, they are also integral to other functions. Where would reproduction (R) be without cilia in fallopian tubes, flagellated sperm, and well-muscled uteruses? Sensitivity (S) to sound requires modulation and amplification by small muscles and electroactive molecular motors inside the ear, and for many animals, moving their external ears to better locate and capture sounds is critical to their survival. Sensitivity to tastes, smells, and sights also benefits from movement. Growth (G) requires cell division many times over, where actin and myosin working in concert repeatedly pinch one cell into two. Respiration (R) not only involves muscles to move the chest wall and diaphragm but also cilia in the nose and trachea to keep the airways clean. Excretion (E) depends on peristalsis in the ureters and the sphincters that control the bladder. Nutrition (N) is aided by smooth muscles encircling the entire digestive system.

In the introduction I asked, "Are you happy with your weight? Strength? Physique? Blood pressure? Blood glucose level? Mental and physical endurance? Sleep pattern? Furthermore, do you want a long and active life?" After I built up courage, I answered the questions myself. Happy with weight? No, especially not in January. Strength? Carrying a full suitcase upstairs is getting harder to do. Physique? So-so. Blood pressure and blood glucose level? My doctor says they're okay. Mental and physical endurance? Who couldn't use more? Sleep pattern? Terrible. Want a long and active life? Absolutely!

With those answers I felt compelled to more thoroughly practice what I preach in chapter 6 about conditioning. I clearly do not have many fast-twitch muscle fibers, so my athletic endeavors have always tilted toward ones that capitalize on endurance—slow-twitch activities, especially running and cycling. I also have our dog take me for daily walks, and I engage regularly in *power gardening*—mowing, digging, crawling, stooping, climbing, lifting, carrying. By these means, I easily average over 10,000 steps a day, and my resting pulse has been in the 50s since I was a young adult. So from a cardiorespiratory perspective, I am probably in pretty good shape.

Conversely, despite knowing full well that weight training is essential for maintaining muscle mass and thereby maintaining bone density, I have always hated pumping iron. I get bored. At least when I dig a hole in the garden I exercise my trunk and limbs and have an abyss to show for my efforts. For the same reason, I like splitting firewood with wedges and a sledgehammer—total body workout. The problem is that there are a limited number of holes to dig and logs to split on a city lot.

Therefore, to better inform readers, I bit the bullet two years ago and hired a personal trainer. He puts me through my paces twice a week, and I do resistance training on my own one more time with guidance and encouragement from a smartphone app or YouTube video. I can't yet say that performing resistance exercises has changed my life, but I feel good the rest of the day, look forward to the next session, and know that I am being proactive regarding my health. That includes watching more carefully what and how much I eat.

Enough about me. I sincerely hope that learning about muscle has encouraged you to up your game and improve your fitness and overall health. You have the groundwork in place, and there is plenty of help available. Your newly gained knowledge also informs your life more broadly, from the history of science and the melding of art and anatomy to popular culture and stupefying technological advances on the horizon. The long and short of it is that muscle is a moving reflection of life in all its beautiful complexity. Make the most of yours.

Acknowledgments

I KNEW THAT I WAS GOING TO LIKE MY NEW AGENT, JOËLLE Delbourgo, when she responded positively within a few days to my inquiry regarding representation for a book on muscle. After that, it just got better. An experienced pro in every respect, Joëlle has been my ardent fan and advocate of *Muscle* and has wisely and calmly guided me through several stresses of manuscript preparation and acceptance.

As with my previous book, *Bones, Inside and Out*, the staff at W. W. Norton has been instrumental in ensuring that muscle's story unfolds coherently and clearly. To this end editor Drew Weitman deserves credit for sometimes advising "forest" when I was seeing "trees." She encouraged me, successfully, to rearrange the order of presentation, condense some material, and eradicate any last vestiges of techno jargon. As a result of Drew's keen editorial eye, readers will greatly, although unknowingly, benefit. To the rest of the staff at W. W. Norton, thank you for making *Muscle* visually appealing and ensuring that learners will find it attractive.

Others also contributed to the story's clarity, accuracy, and visual appeal. Vernon Tolo, Molly Niven, and David Niven graciously, even if silently grimacing, read early drafts of the entire manuscript and made helpful suggestions. Others read portions related to their areas of expertise and provided valued advice. These included Jim Weiss (cardiology), Agnes Brooker (anesthesia), Alex Jamal (physical conditioning, bodybuilding, and nutrition), Bill Bowersock (singing), and Francy Shu (electromyography). Carrie Cantor later added her line-editor's discerning eye and ensured that the text actually said what I wanted to express, and succinctly so. Then Nancy Green, copyeditor extraordinaire, scrutinized every mark and space (approaching half a million) to perfect the manuscript's clarity, grammar, and punctuation. Similarly, graphic designer Elizabeth Conley prepared all the images for reproduction and ensured that even imperfectly preserved historic photographs would contribute to the story.

Impossible to pinpoint, but paramount to the book's final form, are the thousands of patients who have entrusted the care of their musculoskeletal problems to me. Their ailments, recoveries, and observations continue to teach me a great deal, and they have motivated me to share my enthusiasm and awe for this amazing tissue. Likewise, I am grateful to the hundreds of hand surgery fellows, orthopedic and plastic surgery residents, and medical students whom I have had the privilege to teach and who have also been my teachers. I further acknowledge my bona fide teachers, from junior high school on, who, along with camp counselors, introduced me to the study of the natural world and stimulated my curiosity.

Finally, I remained stunned by the power of the Internet, which provided me instant access to the world's literature on life sciences as well as up-to-date information on popular culture. Electronic access monumentally enhanced the ease, depth, and breadth of my background reading.

Supplemental Viewing

MOTION PHOTOGRAPHY AND COMPUTER ANIMATIONS CAN enhance the understanding of muscle movement in ways not fully possible with words and static images. The following video clips are accessible on YouTube by searching for the listed terms. To watch these clips in their entirety will take less than 30 minutes, but I'm guessing that the visual enlightenment will lure you to browse the related offerings that YouTube recommends and learn even more about the world's best motor.

lamellipodia dynamics
growth cones turning and actin dynamics
cytoplasmic streaming: Elodea under the microscope
muscle structure and function animation by Drew Barry
transoral incisionless fundoplication
The Visible Human Project—Male
The Muscle Song (Memorize Your Anatomy)
hungry woodpecker sticks out tongue
Javier Sotomayor high jump world record
kinesin protein walking on microtubule
ATP synthase, Graham Johnson
seven power suits
targeted muscle reinnervation patient
cyborg stingray made of rat muscles and gold

Selected Bibliography

Chapter 1: Discovery and Description

Britannica. "Connective Tissue." Accessed October 18, 2021. https://www.britannica.com/science/skeleton/Connective-tissue.

Burgoon, Judee K. "Microexpressions Are Not the Best Way to Catch a Liar." *Frontiers in Psychology* 9 (2018): 1672.

De Havas, Jack, Hiroaki Gomi, and Patrick Haggard. "Experimental Investigations of Control Principles of Involuntary Movement: A Comprehensive Review of the Kohnstamm Phenomenon." *Brain Research* 235, no. 7 (2017): 1953–97.

Eckman, P. "Facial Expressions of Emotion: An Old Controversy and New Findings." *Philosophical Transactions of the Royal Society B Biological* 335, no. 1273 (1992): 63–69.

Harrison, R. J., and E. J. Field. *Anatomical Terms: Their Origins and Derivation*. Cambridge: W. Heffer and Son, 1947.

Hendriks, I. F., D. A. Zhuravlev, J. G. Govill, F. Boer, I. V. Baivoronskii, P. C. W. Hogendoorn, and M. C. DeRuiter. "Nikolay Ivanovich Pirogov (1810–1881): Anatomical Research to Develop Surgery." *Clinical Anatomy* 33, no. 5 (2020): 714–30.

Hilloowala, Rumy. "Michelangelo: Anatomy and Its Implication in His Art." *Vesalius* 15, no. 1 (2009): 19–24.

Ivanenko, Y. P., W. G. Wright, V. S. Gurfinkel, F. Horak, and P. Cordo. "Interaction of Involuntary Post-Contraction Activity with Locomotor Movements." *Experimental Brain Research* 169, no. 2 (2006): 255–60.

Kemp, Martin. "Style and Non-Style in Anatomical Illustration: From Renaissance Humanism to Henry Gray." *Journal of Anatomy* 216, no. 2 (2010): 192–208.

Kohnstamm, O. "Demonstration einer Katatonieartigen Erscheinung beim Gesunden." *Neurologie Zentralblatt* 34 (1915): 290–91.

Lee, Se-Jin, Adam Lehar, Jessica U. Meir, Christina Koch, Andrew Morgan, Lara E. Warren, Renata Rydzik, et al. "Targeting Myostatin/Activin A Protects Against Skeletal Muscle and Bone Loss during

Spaceflight." *Proceedings of the National Academy of Sciences* 117, no. 38 (2020): 23942–51.

Longo, Aldo F., Carlos R. Siffredi, Marcelo L. Cardey, Gustavo D. Aquilino, and Néstor A. Lentini. "Age of Peak Performance in Olympic Sports: A Comparative Research Among Disciplines." *Journal of Human Sport and Exercise* 11, no. 1 (2016): 31–41.

Macalister, A. "Observations on Muscular Anomalies in the Human Anatomy. Third Series, with a Catalogue of the Principal Muscular Variations Hitherto Published." *Transactions of the Royal Irish Academy of Science* 25 (1875): 1–130.

Musil, Vladimir, Zdenek Suchomel, Petra Malinova, Josef Stingl, Martin Vlcek, and Marek Vacha. "The History of Latin Terminology of Human Skeletal Muscles (From Vesalius to the Present)." *Surgical and Radiologic Anatomy* 37 (2015): 33–41.

Shelbourn, Carolyn. "Bringing the Skeletons Out of the Closet: The Law and Human Remains in Art, Archaeology and Museum Collections." *Art Antiquity and Law Journal* 11 (2006): 179–98.

Spira, Anthony, Martin Postle, and Paul Bonaventura. *George Stubbs: "All Done from Nature."* London: Paul Holberton Publishing, 2019.

Spitzer, Victor M., and David G. Whitlock. *National Library of Medicine Atlas of the Visible Human Male: Reverse Engineering of the Human Body.* Burlington, MA: Jones and Bartlett Learning, 1997.

Starr, Michelle. "Some People Can Make a Roaring Sound in Their Ears Just by Tensing a Muscle." Accessed October 18, 2021. https://www.sciencealert.com/some-people-can-make-a-roaring-sound-in-your-ears-just-by-tensing-a-muscle.

Vasari, G. *The Lives of the Most Excellent Painters, Sculptors, and Architects.* Florence, 1550 and 1568. *The Lives of the Artists.* Translated by Julia Conway Bondanella. Oxford World's Classics. Oxford: Oxford University Press, 2008.

Voloshin, I., and P. M. Bernini. "Nickolay Ivanovich Pirogoff. Innovative Scientist and Clinician." *Spine* 23, no. 19 (1976): 2143–46.

Woltmann, Alfred, and Karl Woermann. *History of Painting*, vol. 2, *The Painting of the Renascence.* Translated by Clara Bell. New York: Dodd, Mead, 1885.

Wood J. "Variations in Human Myology." *Proceedings of the Royal Society of London* 16 (1868): 483–525.

Chapter 2: Molecular Magic

Akasaki, Y., N. Ouchi, Y. Izumiya, B. Bernardo, N. LeBrasseur, and K. Walsh. "Glycolytic Fast-Twitch Muscle Fiber Restoration Counters Adverse Age-Related Changes in Body Composition and Metabolism." *Aging Cell* 13 (2013): 80–91.

Al-Khayat, Hind A. "Three-Dimensional Structure of the Human Myosin

Thick Filament: Clinical Implications." *Global Cardiology and Scientific Practice* 2013, no. 3 (2013): 280–302.

Austin City College. "Muscle Cell Anatomy and Function." Accessed October 16, 2021. https://www.austincc.edu/sziser/Biol%202404/2404LecNotes/2404LNExII/Muscle%20Physiology.pdf.

Boland, Mike, Lovedeep Kaur, Feng Ming Chian, and Thierry Astruc. "Muscle Proteins." Accessed October 16, 2021. https://hal.archives-ouvertes.fr/hal-02000883/document.

Cohen, Joe. "What Does Myostatin Inhibition Do? + Risks & Side Effects." Accessed October 16, 2021. https://selfhacked.com/blog/myostatin-inhibition/.

Čolović, Mirjana B, Danijela Z. Krstić, Tamara D. Lazarević-Pašti, Aleksandra M. Bondžić, and Vesna M. Vasić. "Acetylcholinesterase Inhibitors: Pharmacology and Toxicology." *Current Neuropharmacology* 11, no. 3 (2013) 315–35.

Hamilton, Jon. "Scientists Sent Mighty Mice to Space to Improve Treatments Back on Earth." Accessed October 16, 2021. https://www.npr.org/sections/health-shots/2020/01/16/796316186/scientists-sent-mighty-mice-to-space-to-improve-treatments-back-on-earth.

Hartman, M. Amanda, and James A. Spudich. "The Myosin Superfamily at a Glance." *Journal of Cell Science* 128, no. 11 (2015): 2009–19.

Narici, M. V., and N. Maffulli. "Sarcopenia: Characteristics, Mechanisms and Functional Significance." *British Medical Bulletin* 95 (2010): 139–59.

Nigam, P. K., and Anjana Nigam. "Botulinum Toxin." *Indian Journal of Dermatology* 55, no. 1 (2010): 8–14.

Orizio, Claudio. "Muscle Sound: Bases for the Introduction of a Mechanomyographic Signal in Muscle Studies." *Critical Reviews in Biomedical Engineering* 21, no. 3 (1993): 210–43.

Oster, Gerald. "Muscle Sounds." *Scientific American* 250, no. 3 (1984): 108–15.

Ross, T. "Myostatin Inhibitors—Do They Work? Is There Another Way to Do It?" Accessed October 16, 2021. https://www.researchedsupplements.com/myostatin-inhibitors.

Stanford University. "The History of Muscles." Accessed October 16, 2021. https://web.stanford.edu/class/history13/earlysciencelab/body/musclespages/muscles.html.

Szent-Györgyi, Andrew G. "The Early History of the Biochemistry of Muscle Contraction." *Journal of General Physiology* 123, no. 6 (2004): 631–41.

White, T. A., and N. K. LeBrasseur. "Myostatin and Sarcopenia: Opportunities and Challenges—A Mini-Review." *Gerontology* 60 (2014): 289–93.

Chapter 3: Skeletal Muscle

Adams, Valerie, Bernice Mathisen, Surinder Baines, Cathy Lazarus, and Robin Callister. "A Systematic Review and Meta-Analysis of Measurements of Tongue and Hand Strength and Endurance Using the Iowa Oral Performance Instrument (IOPI)." *Dysphagia* 28, no. 3 (2013): 350–69.

Dikmen, Ebrar. "Embodied Cognition: Change Your Mental Level with Your Body." Accessed October 18, 2021. https://mozartcultures.com/en/uyelik/.

Goss, C. M. "On Movement of Muscles by Galen of Pergamon." *American Journal of Anatomy* 123, no. 1 (1968): 1–26.

Hung, Iris W., and Aparna A. Labroo. "From Firm Muscles to Firm Willpower: Understanding the Role of Embodied Cognition in Self-Regulation." *Journal of Consumer Research* 37, no. 6 (2011): 1046–64.

Janssen, Ian, Steven B. Heymsfield, and Z. M. Wang. "Skeletal Muscle Mass and Distribution in 468 Men and Women Aged 18–88 Yr." *Journal of Applied Physiology* 89, no. 1 (1985): 81–88.

Kayalioglu, Gulgun, Baris Altay, Feray Gulec Uyaroglu, Fikret Bademkiran, Burhanettin Uludag, and Cumhur Ertekin. "Morphology and Innervation of the Human Cremaster Muscle in Relation to Its Function." *Anatomical Record* 29, no. 7 (2008): 790–96.

Kemp, Martin. "Style and Non-Style in Anatomical Illustration: From Renaissance Humanism to Henry Gray." *Journal of Anatomy* 216, no. 2 (2010): 192–208.

Kurivan, Rebecca. "Body Composition Techniques." *Indian Journal of Medical Research* 148, no. 5 (2018): 648–58.

Shepherd, John, Bennett Ng, Markus Sommer, and Steven B. Heymsfield. "Body Composition by DXA." *Bone* 104 (2017): 101–5.

Tuthill, John C., Eiman Azim, and Ian Waterman. "Proprioception." *Current Biology* 28, no. 5 (2018): R194–R203.

Vesalius, Andreas. *De Humani Corporis Fabrica Libri Septum (On the Fabric of the Human Body)*, translated by William Frank Richardson. San Francisco: Norman Publishing, 1999.

Chapter 4: Smooth Muscle

Ackerknecht, E. H. "The History of the Discovery of the Vegetative (Autonomic) Nervous System." *Medical History* 18, no. 1 (1974): 1–8.

Agarwal, Pawan, Shabbir Husain, Sudesh Wankhede, and D. Sharma. "Rectus Abdominis Detrusor Myoplasty (RADM) for Acontractile/Hypocontractile Bladder in Spinal Cord Injury Patients: Preliminary Report." *Journal of Plastic, Reconstructive & Aesthetic Surgery* 71, no. 5 (2018): 736–42.

Amrani, Yassine, and Reynold A. Panettieri. "Airway Smooth Muscle:

Contraction and Beyond." *International Journal of Biochemistry and Cell Biology* 35, no. 3 (2003): 272–76.

Barišić, Goran, and Zoran Krivokapić. "Adynamic and Dynamic Muscle Transposition Techniques for Anal Incontinence." *Gastroenterology Report* 2, no. 2 (2014): 98–105.

Benson, Herbert, John W. Lehmann, M. S. Malhotra, Ralph F. Goldman, Jeffrey Hopkins, and Mark D. Epstein. "Body Temperature Changes during the Practice of G Tum-mo Yoga." *Nature* 295 (1982): 234–36.

Bharadwaj, Shishira, Parul Tandon, Tushar D. Gohel, Jill Brown, Ezra Steiger, Donald F. Kirby, Ajai Khanna, et al. "Current Status of Intestinal and Multivisceral Transplantation." *Gasteroenterology Report (Oxford)* 5, no. 1 (2017): 20–28.

Bianchi, A. "Intestinal Loop Lengthening—A Technique for Increasing Small Intestinal Length." *Journal of Pediatric Surgery* 15, no. 2 (1980): 145–51.

Bornemeier, Walter C. "Sphincter Protecting Hemorrhoidectomy." *American Journal of Proctology* 11 (1960): 48–52.

Bourne, L. E., C. P. Wheeler-Jones, and I. R. Orriss. "Regulation of Mineralisation in Bone and Vascular Tissue: A Comprehensive Review." *Journal of Endocrinology* 248, no. 2 (2021): R51–R65.

Bubenik, George A. "Why Do Humans Get 'Goosebumps' When They Are Cold, Or under Other Circumstances?" Accessed October 17, 2021. https://open.oregonstate.education/aandp/chapter/10-7-smooth-muscle-tissue/.

Cleveland Clinic. "For the First Time in North America, a Women Gives Birth after Uterus Transplant from a Deceased Donor." Accessed October 17, 2021. https://health.clevelandclinic.org/for-the-first-time-in-north-america-woman-gives-birth-after-uterus-transplant-from-deceased-donor/.

Dalziel J. E., N. J. Spencer, and W. Young. "Microbial Signalling in Colonic Motility." *International Journal of Biochemistry & Cell Biology* 134 (2021): 105963.

Deora, Surender. "The Story of 'STENT': From Noun to Verb." *Indian Heart Journal* 68, no. 2 (2016): 235–37.

Dunn, P. "John Braxton Hicks (1893–97) and Painless Uterine Contractions." *Archives of Diseases in Children. Fetal and Neonatal Edition* 81, no. 2 (1999): F157–F158.

Grootaert, Mandy O. J., and Martin R Bennett. "Smooth Muscle Cells in Atherosclerosis: Time for a Reassessment." *Cardiovascular Research* 117, no. 11 (2021): 2326–39.

Grubb, Søren, Changsi Cai, Bjørn O. Hald, Lila Khennouf, Reena Prity Murmu, Aske G. K. Jensen, Jonas Fordsmann, et al. "Precapillary Sphincters Maintain Perfusion in the Cerebral Cortex." *Nature Communications* 11, no. 1 (2020): 395.

Gutowski, Piotr, Shawn M. Gage, Malgorzata Guziewicz, Marek Ilzecki,

Arkadiusz Kazimierczak, Robert D. Kirkton, Laura E. Niklason, et al. "Arterial Reconstruction with Human Bioengineered Acellular Blood Vessels in Patients with Peripheral Arterial Disease." *Journal of Vascular Surgery* 72, no. 4 (2020): 1247–58.

Harryman, William L., Kendra D. Marr, Daniel Hernandez-Cortes, Raymond B. Nagle, Joe G. N. Garcia, and Anne E. Cress. "Cohesive Cancer Invasion of the Biophysical Barrier of Smooth Muscle." *Cancer Metastasis Review* 40, no. 1 (2021): 205–19.

Hicks J. B. "On the Contractions of the Uterus Throughout Pregnancy: Their Physiological Effects and Their Value in the Diagnosis of Pregnancy." *Transactions of the Obstetrical Society of London* 13 (1871): 216–31.

Hur, Christine, Jenna Rehmer, Rebecca Flyckt, and Tommaso Falcone. "Uterine Factor Infertility: A Clinical Review." *Clinical Obstetrics and Gynecology* 62, no. 2 (2019): 257–70.

Ilardo, Melissa A., Ida Moltke, Thorfinn S. Korneliussen, Jade Cheng, Aaron J. Stern, Fernando Racimo, Peter de Barros Damgaard, et al. "Physiological and Genetic Adaptations to Diving in Sea Nomads." *Cell* 173, no. 3 (2018): 569–80.e15.

Jones, B. P., N. J. Williams, S. Saso, M.-Y. Thum, I. Quiroga, J. Yazbek, S. Wilkinson, et al. "Uterine Transplantation in Transgender Women." *British Journal of Obstetrics and Gynaecology* 126, no. 2 (2019): 152–56.

Jones, T. W. "Discovery That Veins of the Bat's Wing (Which Are Furnished with Valves) Are Endowed with Rhythmical Contractility and That the Onward Flow of Blood Is Accelerated by Each Contraction." *Philosophical Transactions of the Royal Society of London* 142 (1852): 131–36.

Katayama, Rafael C., Fernando A. M. Herbella, Francisco Schlottmann, and P. Marco Fisichella. "Lessons Learned from the History of Fundoplication." *SN Comprehensive Clinical Medicine* 2 (2020): 775–81.

Kegel, A. H. "The Nonsurgical Treatment of Genital Relaxation; Use of the Perineometer as an Aid in Restoring Anatomic and Functional Structure." *Annals of Western Medicine and Surgery* 2, no. 5 (1948): 213–16.

Kesseli, S., and D. Sudan. "Small Bowel Transplantation." *Surgical Clinics of North America* 99, no. 1 (2019): 103–16.

Kirkton, Robert D., Maribel Santiago-Maysonet, Jeffrey H. Lawson, William E. Tente, Shannon L. M. Dahl, Laura E. Niklason, and Heather L. Prichard. "Bioengineered Human Acellular Vessels Recellularize and Evolve into Living Blood Vessels after Human Implantation." *Science Translational Medicine* 11, no. 485 (2019): eaau6934.

Kozhevnikov, Maria, James Elliott, Jennifer Shephard, and Klaus Gramann. "Neurocognitive and Somatic Components of Temperature Increases during g-Tummo Meditation: Legend and Reality." *PLOS One* 8, no. 3 (2013): e58244.

Krishnan, Jerry A., and Aliya N. Husain. "One Step Forward, Two Steps

Back: Bronchial Thermoplasty for Asthma." *American Journal of Respiratory and Critical Care Medicine* 203, no. 2 (2021): 153–54.

Lehur, Paul-Antoine, Shane McNevin, Steen Buntzen, Anders F. Mellgren, Soeren Laurberg, and Robert D. Madoff. "Magnetic Anal Sphincter Augmentation for the Treatment of Fecal Incontinence: A Preliminary Report from a Feasibility Study." *Diseases of the Colon and Rectum* 53, no. 12 (2010): 1604–10.

Luo, Jiesi, Yuyao Lin, Xiangyu Shi, Guangxin Li, Mehmet H. Kural, Christopher W. Anderson, Matthew W. Ellis, et al. "Xenogeneic-Free Generation of Vascular Smooth Muscle Cells from Human Induced Pluripotent Stem Cells for Vascular Tissue Engineering." *Acta Biomaterialia* 119 (2021): 155–68.

Magalhaes, Renata S., J. Koudy Williams, Kyung W. Yoo, James J. Yoo, and Anthony Atala. "A Tissue-Engineered Uterus Supports Live Birth in Rabbits." *Nature Biotechnology* 38 (2020): 1280–87.

Matsumoto, C. S., S. Subramanian, and T. M. Fishbein. "Adult Intestinal Transplantation." *Gasteroenterology Clinics of North America* 47, no. 2 (2018): 341–54.

Mookerjee, Vikram G., and Daniel Kwan. "Uterus Transplantation as a Fertility Option in Transgender Health Care." *International Journal of Transgender Health* 21, no. 2 (2021): 122–24.

Nguyen, Jennifer V. "The Biology, Structure, and Function of Eyebrow Hair." *Journal of Drugs in Dermatology* 13, no. 1 Suppl. (2014): s12–s16.

Nyangoh, Timoh K., D. Moszkowicz, M. Creze, M. Zaitouna, M. Felber, C. Lebacle, D. Diallo, et al. "The Male External Sphincter Is Autonomically Innervated." *Clinical Anatomy* 34, no. 2 (2021): 263–71.

Oregon State University. "Smooth Muscle Tissue." Accessed October 17, 2021. https://open.oregonstate.education/aandp/chapter/10-7-smooth-muscle-tissue./

Oshiro, Takuma, Ryu Kimura, Keiichiro Izumi, Asuka Ashikari, Seiichi Saito, and Minoru Miyazato. "Changes in Urethral Smooth Muscle and External Urethral Sphincter Function with Age in Rats." *Physiological Reports* 8, no. 24 (2021): e14643.

Panneton, W. Michael. "The Mammalian Diving Response: An Enigmatic Reflex to Preserve Life?" *Physiology (Bethesda)* 28, no. 5 (2013): 284–97.

Paus, I., I. Burgoa, C. I. Platt, T. Griffiths, E. Poblet, and A. Izeta. "Biology of the Eyelash Hair Follicle: An Enigma in Plain Sight." *British Journal of Dermatology* 174, no. 2 (2016): 741–52.

Pollard, Stephen. "Small Bowel Transplantation." *Clinics in Colon Rectal Surgery* 17, no. 2 (2004): 119–24.

Sarveazad, Arash, Asrin Babahajian, Naser Amini, Jebreil Shamseddin, and Mahmoud Yousefifard. "Posterior Tibial Nerve Stimulation in Fecal Incontinence: A Systematic Review and Meta-Analysis." *Basic and Clinical Neuroscience* 10, no. 5 (2019): 419–31.

Smith, Edwin A., Jonathan D. Kaye, John Y. Lee, Andrew J. Kirsch, and Joseph K. Williams. "Use of Rectus Abdominis Muscle Flap as Adjunct to Bladder Neck Closure in Patients with Neurogenic Incontinence: Preliminary Experience." *Journal of Urology* 183, no. 4 (2010): 1556–60.

Tzvetanov, Ivo G., Kiara A. Tulla, Giuseppe D'Amico, and Enrico Benedetti. "Living Donor Intestinal Transplantation." *Gastroenterology Clinics of North America* 47, no. 2 (2018): 369–80.

Wan, Juyi, Xiaolin Zhong, Zhiwei Xu, Da Gong, Diankun Li, Zhifei Xin, Xiaolong Ma, et al. "A Decellularized Porcine Pulmonary Valved Conduit Embedded with Gelatin." *Artificial Organs* 45, no. 9 (2021): 1068–82.

Whitehead A. K., A. P. Erwin, and X. Yue. "Nicotine and Vascular Dysfunction." *Acta Physiologica (Oxford)* 17 (2021): e13631.

Witcombe, Brian, and Dan Meyer. "Sword Swallowing and Its Side Effects." *British Medical Journal* 333, no. 7582 (2006): 1285–87.

Yuan, Cheng, Lihua Ni, Changjiang Zhang, Xiaorong Hu, and Xiaoyan Wu. "Vascular Calcification: New Insights into Endothelial Cells." *Microvascular Research* 134 (2021): 104–5.

Zhao, Jian, Zhaoyu Liu, and Zhihui Chang. "Osteogenic Differentiation and Calcification of Human Aortic Smooth Muscle Cells Is Induced by the RCN2/STAT3/miR-155-5p Feedback Loop." *Vascular Pharmacology* 136 (2021): 106821.

Chapter 5: Cardiac Muscle

American Heart Association. "The Past, Present and Future of the Device Keeping Alive Carew, Thousands of HF Patients." Accessed October 15, 2021. https://www.heart.org/en/news/2018/06/13/the-past-present-and-future-of-the-device-keeping-alive-carew-thousands-of-hf-patients.

Aranki, Sary. "Coronary Artery Bypass Graft Surgery: Graft Choices." Accessed October 21, 2021. https://www.uptodate.com/contents/coronary-artery-bypass-graft-surgery-graft-choices.

Bakhtiyar, Syed Shahyan, Elizabeth L. Godfrey, Shayan Ahmed, Harveen Lamba, Jeffrey Morgan, Gabriel Loor, Andrew Civitello, et al. "Survival on the Heart Transplant Waiting List." *JAMA Cardiology* 5, no. 11 (2020): 1227–35.

Barron, S. L. "Development of the Electrocardiograph in Great Britain." *British Medical Journal* 1, no. 4655 (1950): 720–25.

Becnel, Miriam, and Selim R. Krim. "Left Ventricular Assist Devices in the Treatment of Advanced Heart Failure." *Journal of the American Academy of Physician Assistants* 32, no. 5 (2019): 41–46.

Braunwald, Eugene. "Cell-Based Therapy in Cardiac Regeneration: An Overview." *Circulation Research* 123, no. 2 (2018): 132–37.

Brenner, Paolo, and Maks Mihalj. "Update and Breakthrough in Cardiac

Xenotransplantation." *Current Opinion in Organ Transplantation* 25, no. 3 (2020): 261–67.

Curfman. G. "Stem Cell Therapy for Heart Failure: An Unfulfilled Promise?" *JAMA* 321 (2019): 1186–87.

Domingues, José Sérgio, Marcos de Paula Vale, and Marcos Pinotti Barbosa. "Partial Left Ventriculectomy: Have Well-Succeeded Cases and Innovations in the Procedure Been Observed in the Last 12 Years?" *Brazilian Journal of Cardiovascular Surgery* 30, no. 5 (2015): 579–85.

Duan, Dongsheng, and Jerry R. Mendell, editors. *Muscle Gene Therapy*. 2nd ed. New York: Springer, 2019.

Edwards, Jena E., Elizabeth Hiltz, Franziska Broell, Peter G. Bushnell, Steven E. Campana, Jørgen S. Christiansen, Brynn M. Devine, et al. "Advancing Research for the Management of Long-Lived Species: A Case Study on the Greenland Shark." *Frontiers in Marine Science* 6 (2019): 87.

Fahlman, Andreas, Bruno Cozzi, Mercey Manley, Sandra Jabas, Marek Malik, Ashley Blawas, and Vincent M. Janik. "Conditioned Variation in Heart Rate during Static Breath-Holds in the Bottlenose Dolphin (*Tursiops truncatus*)." *Frontiers in Physiology* 11 (2020): 604018.

Fernández-Ruiz, Irene. "Breakthrough in Heart Xenotransplantation." *Nature Reviews Cardiology* 16, no. 2 (2019): 69.

Gao, L., and J. J. Zhang. "Efficient Protocols for Fabricating a Large Human Cardiac Muscle Patch from Human Induced Pluripotent Stem Cells." *Methods in Molecular Biology* 2158 (2021): 187–97.

Ghiroldi, A., M. Piccoli, G. Ciconte, C. Pappone, and L. Anastasia. "Regenerating the Human Heart: Direct Reprogramming Strategies and Their Current Limitations." *Basic Research in Cardiology* 112, no. 6 (2017): 68.

Goff, Z. D., A. B. Kichura, J. T. Chibnall, and P. J. Hauptman. "A Survey of Unregulated Direct-to-Consumer Treatment Centers Providing Stem Cells for Patients with Heart Failure." *JAMA Internal Medicine* 177, no. 9 (2017): 1387–88.

Goldbogen, J. A., D. E. Cade, J. Calambokidis, M. F. Czapanskiy, J. Fahlbusch, A. S. Friedlaender, W. T. Gough, et al. "Extreme Bradycardia and Tachycardia in the World's Largest Animal." *Proceedings of the National Academy of Sciences* 116, no. 50 (2019): 25329–32.

Han, Jooli, and Dennis R. Trumble. "Cardiac Assist Devices: Early Concepts, Current Technologies, and Future Innovations." *Bioengineering (Basel)* 6, no. 1 (2019): 18.

Harris, K. M., S. Mackey-Bojack, M. Bennett, D. Nwaudo, E. Duncanson, and B. J. Maron. "Sudden Unexpected Death Due to Myocarditis in Young People." *American Journal of Cardiology* 143 (2021): 131–34.

Hayward, M. "Dynamic Cardiomyoplasty: Time to Wrap It Up?" *Heart* 82, no. 3 (1999): 263–64.

Hetzer, Roland, Mariano Francisco del Maria Javier, Frank Wagner, Matthias Loebe, and Eva Maria Javier Delmo. "Organ-Saving Surgical Alternatives to Treatment of Heart Failure." *Cardiovascular Diagnosis and Therapy* 11, no. 1 (2021): 213–25.

Huang, Ke, Emily W. Ozpinar, Teng Su, Junnan Tang, Deliang Shen, Li Qiao, Shiqi Hu, et al. "An Off-the-Shelf Artificial Cardiac Patch Improves Cardiac Repair after Myocardial Infarction in Rats and Pigs." *Science Translational Medicine* 12, no. 538 (2020): eaat9683.

Kaye, Alan D., Allyson L. Spence, Mariah Mayerle, Nitish Sardana, Claire M. Clay, Matthew R. Eng, Markus M. Luedi, et al. "Impact of COVID-19 Infection on the Cardiovascular System: An Evidence-Based Analysis of Risk Factors and Outcomes." *Best Practices and Research in Clinical Anaesthesiology* 35, no. 3 (2021): 437–48.

Leier, Carl V. "Editorial Comment Cardiomyoplasty: Is It Time to Wrap It Up?" *Journal of the American College of Cardiology* 28, no. 5 (1996): 1181–82.

Li, W., S. A. Su, J. Chen, H. Ma, and M. Xiang. "Emerging Roles of Fibroblasts in Cardiovascular Calcification." *Journal of Cellular and Molecular Medicine* 25, no. 4 (2021): 1808–16.

Lim, Gregory B. "An Acellular Artificial Cardiac Patch for Myocardial Repair." *Nature Reviews Cardiology* 17 (2020): 220.

Long, Ashleigh, and Paul Mahoney. "Use of Mitral Clip to Target Obstructive SAM in Severe Diffuse-Type Hypertrophic Cardiomyopathy: Case Report and Review of Literature." *Journal of Invasive Cardiology* 32, no. 9 (2020): E228–E232d.

Maron, B. J., T. S. Haas, A. Ahluwalia, C. J. Murphy, and R. F. Garberich. "Demographics and Epidemiology of Sudden Deaths in Young Competitive Athletes: From the United States National Registry." *American Journal of Medicine* 129, no. 11 (2016): 1170–77.

Martinez, W. Matthew, Andrew M. Tucker, O. Josh Bloom, Gary Green, John P. DiFiori, Gary Solomon, Dermot Phelan, et al. "Prevalence of Inflammatory Heart Disease Among Professional Athletes with Prior COVID-19 Infection Who Received Systematic Return-to-Play Cardiac Screening." *JAMA Cardiology* 6, no. 7 (2021): 745–52.

McGregor, Christopher G. A., and W. Byrne Guerand. "Porcine Human Heart Transplantation: Is Clinical Application Now Appropriate?" *Journal of Immunology Research* 2017 (2017): 2534653.

Mei, Xuan, and Ke Cheng. "Recent Development in Therapeutic Cardiac Patches." *Frontiers in Cardiovascular Medicine* 27, no. 7 (2020): 610364.

Meier, Raphael P. H., Alban Longchamp, Muhammad Mohiuddin, Oriol Manuel, Georgios Vrakas, Daniel G. Maluf, Leo H. Buhler, Yannick D. Muller, and Manuel Pascual. "Recent Progress and Remaining Hurdles Toward Clinical Xenotransplantation." *Xenotransplantation* (March 23, 2021): e12681.

National Cancer Institute. "Matters of the Heart: Why Are Cardiac

Tumors So Rare?" Accessed October 15, 2021. https://www.cancer.gov/types/metastatic-cancer/research/cardiac-tumors.

Piccolino, Marco. "Animal Electricity and the Birth of Electrophysiology: The Legacy of Luigi Galvani." *Brain Research Bulletin* 46, no. 5 (1998): 381–407.

Pierson III, Richard N., Jay A. Fishman, Gregory D. Lewis, David A. D'Alessandro, Margaret R. Connolly, Lars Burdorf, Joren C. Madsen, and Agnes M. Azimzadeh. "Progress Toward Cardiac Xenotransplantation." *Circulation* 142 (2020): 1389–98.

Puntmann, Valentina O., M. Ludovica Carerj, Imke Wieters, Masia Fahim, Christophe Arendt, Jedrzej Hoffmann, Anastasia Shchendrygina, et al. "Outcomes of Cardiovascular Magnetic Resonance Imaging in Patients Recently Recovered from Coronavirus Disease 2019 (COVID-19)." *JAMA Cardiology* 5, no. 11 (2020): 1265–73.

Quick, Nicola J., William R. Cioffi, Jeanne M. Shearer, Andreas Fahlman, and Andrew J. Read. "Extreme Diving in Mammals: First Estimates of Behavioural Aerobic Dive Limits in Cuvier's Beaked Whales." *Journal of Experimental Biology* 23, no. 223 (Pt 18) (2020): jeb 222109.

Rajpal, Saurabh, Matthew S. Tong, James Borchers, Karolina M. Zareba, Timothy P. Obarski, Orlando P. Simonetti, and Curt J. Daniels. "Cardiovascular Magnetic Resonance Findings in Competitive Athletes Recovering from COVID-19 Infection." *JAMA Cardiology* 6, no. 1 (2021): 116–18.

Rastegar, Hassan, Griffin Boll, Ethan J. Rowin, Noreen Dolan, Catherine Carroll, James E. Udelson, Wendy Wang, et al. "Results of Surgical Septal Myectomy for Obstructive Hypertrophic Cardiomyopathy: The Tufts Experience." *Annals of Cardiothoracic Surgery* 6 (2017): 353–63.

Retraction Watch. "Anversa Cardiac Stem Cell Lab Earns 13 Retractions." Accessed October 15, 2021. https://retractionwatch.com/2018/12/13/anversa-cardiac-stem-cell-lab-earns-13-retractions.

Schagatay, Erika, and Boris Holm. "Effects of Water and Ambient Air Temperatures on Human Diving Bradycardia." *European Journal of Applied Physiology* 73 (1996): 1–6.

Sherrid, Mark V., Daniele Massera, and Daniel G. Swistell. "Surgical Septal Myectomy and Alcohol Ablation: Not Equivalent in Efficacy or Survival." *Journal of the American College of Cardiology* 79, no. 17 (2022): 1656–59.

Streeter, Benjamin W., and Michael E. Davis. "Therapeutic Cardiac Patches for Repairing the Myocardium." *Advances in Experimental Medicine and Biology* 5 (2019): 1–24.

Thompson, Randall C., Adel H. Allam, Guido P. Lombardi, L. Samuel Wann, M. Linda Sutherland, James D. Sutherland, Muhammad Al-Tohamy Soliman, et al. "Atherosclerosis Across 4000 Years of Human History: The Horus Study of Four Ancient Populations." *Lancet* 381, no. 9873 (2013): 1211–22.

Vanderbilt University School of Medicine. "World Leader in Heart Transplants." Accessed October 15, 2021. https://medschool.vanderbilt.edu/vanderbilt-medicine/world-leader-in-heart-transplants/.

Wenger, M. A., B. K. Bagchi, and B. K. Anand. "Experiments in India on "Voluntary" Control of the Heart and Pulse." *Circulation* 24 (1961): 1319–25.

Wilson, Clare. "Myocarditis Is More Common after Covid-19 Infection Than Vaccination." Accessed October 15, 2021. https://www.newscientist.com/article/mg25133462–800-myocarditis-is-more-common-after-covid-19-infection-than-vaccination/#ixzz79P8NxsjA.

Chapter 6: Conditioning

Armstrong, Brock. "Build Strength and Muscle Fast with Occlusion Training." Accessed October 20, 2021. https://www.scientificamerican.com/article/build-strength-and-muscle-fast-with-occlusion-training/.

Bachman, Rachel. "An Olympian, a Failed Drug Test and an Accused Burrito." Accessed October 21, 2021. https://www.wsj.com/articles/olympic-runner-failed-drug-test-burrito-shelby-houlihan-11623867500?mod=hp_major_pos1#cxrecs_s.

Bagheri, Reza, Babak Hooshmand Moghadam, Damoon Ashtary-Larky, Scott C. Forbes, Darren G. Dandow, Andrew J. Galpin, Mozhgan Eskandari, Richard B. Kreider, and Alexie Wong. "Whole Egg vs. Egg White Ingestion during 12 Weeks of Resistance Training in Trained Young Males: A Randomized Controlled Trial." *Journal of Strength and Conditioning Research* 35, no. 2 (2021): 411–19.

Baker, B. S., M. S. Stannard, D. L. Duren, J. L. Cook, and J. P. Stannard. "Does Blood Flow Restriction Therapy in Patients Older Than Age 50 Result in Muscle Hypertrophy, Increased Strength, or Greater Physical Function? A Systematic Review." *Clinical Orthopaedics and Related Research* 478, no. 3 (2020): 593–606.

Berger, Joshua, Oliver Ludwig, Stephan Becker, Marco Backfisch, Wolfgang Kemmler, and Michael Fröhlich. "Effects of an Impulse Frequency Dependent 10-Week Whole-Body Electromyostimulation Training Program on Specific Sport Performance Parameters." *Journal of Sports Science and Medicine* 19, no. 2 (2020): 271–81.

Black, Jonathan. "Charles Atlas: Muscle Man." Accessed October 20, 2021. https://www.smithsonianmag.com/history/charles-atlas-muscle-man-34626921/.

Blocquiaux, Sara, Tatiane Gorski, Evelien Van Roie, Monique Ramaekers, Ruud Van Thienen, Henri Nielens, Christophe Delecluse, Katrien De Bock, and Martine Thomis. "The Effect of Resistance Training, Detraining and Retraining on Muscle Strength and Power, Myofibre Size, Satellite Cells and Myonuclei in Older Men." *Experimental Gerontology* 133 (2020): 110860.

Bowman, Eric N., Rami Elshaar, Heather Milligan, Gregory Jue, Karen Mohr, Patty Brown, Drew M. Watanabe, and Orr Limpisvasti. "Proximal, Distal, and Contralateral Effects of Blood Flow Restriction Training on the Lower Extremities: A Randomized Controlled Trial." *Sports Health* 11, no. 2 (2019): 149–56.

Bowman, Eric N., Rami Elshaar, Heather Milligan, Gregory Jue, Karen Mohr, Patty Brown, Drew M. Watanabe, and Orr Limpisvasti. "Upper-Extremity Blood Flow Restriction: The Proximal, Distal, and Contralateral Effects—A Randomized Controlled Trial." *Journal of Shoulder and Elbow Surgery* 29, no. 6 (2020): 1267–74.

Brandner, C. R., and S. A. Warmington. "Delayed Onset Muscle Soreness and Perceived Exertion after Blood Flow Restriction Exercise." *Journal of Strength and Conditioning Research* 31, no. 11 (2017): 3101–8.

Britannica. "Milo of Croton. Greek Athlete." Accessed October 21, 2021. https://www.britannica.com/biography/Milo-of-Croton.

Calderone, Julia, and Ben Fogelson. "Fact or Fiction? The Tongue Is the Strongest Muscle in the Body." Accessed October 20, 2021. https://www.scientificamerican.com/article/fact-or-fiction-the-tongue-is-the-strongest-muscle-in-the-body/.

Carr, Joshua C., Xin Ye, Matt S. Stock, Michael G. Bemben, and Jason M. DeFreitas. "The Time Course of Cross-Education during Short-Term Isometric Strength Training." *European Journal of Applied Physiology* 119 (2019): 1395–407.

Cohen, Joe. "What Does Myostatin Inhibition Do? + Risks & Side Effects." Accessed October 21, 2021. https://selfhacked.com/blog/myostatin-inhibition/.

Dalleck, Lance C., and Len Kravitz. "The History of Fitness." Accessed October 21, 2021. https://www.unm.edu/~lkravitz/Article%20folder/history.html.

Damas, Felipe, Cleiton A. Libardi, and Carlos Ugrinowitsch. "The Development of Skeletal Muscle Hypertrophy Through Resistance Training: The Role of Muscle Damage and Muscle Protein Synthesis." *European Journal of Applied Physiology* 118, no. 3 (2018): 485–500.

Fair, John D. *Muscletown USA: Bob Hoffman and the Manly Culture of York Barbell.* University Park: Penn State University Press, 1999.

Fleckenstein, Daniel, Olaf Ueberschär, Jan C., Wüstenfeld, Peter Rüdrich, and Bernd Wolfarth. "Effect of Uphill Running on VO_2, Heart Rate and Lactate Accumulation on Lower Body Positive Pressure Treadmills." *Sports (Basel)* 9, no. 4 (2021): 51.

Furtado, Guilherme Eustáquio, Rubens Vinícius Letieri, Adriana Silva-Caldo, Joice C. S. Trombeta, Clara Monteiro, Rafael Nogueira Rodrigues, Ana Vieira-Pedrosa, et al. "Combined Chair-Based Exercises Improve Functional Fitness, Mental Well-Being, Salivary Steroid Balance, and Anti-microbial Activity in Pre-frail Older Women." *Frontiers in Psychology* 12 (2021): 564490.

Garber, Carol Ewing, Bryan Blissmer, Michael R. Deschenes, Barry A. Franklin, Michael J. Lamonte, I-Min Lee, David C. Nieman, David P. Swain, and the American College of Sports Medicine. "American College of Sports Medicine Position Stand. Quantity and Quality of Exercise for Developing and Maintaining Cardiorespiratory, Musculoskeletal, and Neuromotor Fitness in Apparently Healthy Adults: Guidance for Prescribing Exercise." *Medicine and Science in Sports and Exercise* 43, no. 7 (2011): 1334–59.

Government of Canada. "What Happens to Muscles in Space." Accessed October 20, 2021. https://www.asc-csa.gc.ca/eng/sciences/osm/muscles.asp.

Gutierrez, Sara Duarte, Samuel da Silva Aguiar, Lucas Pinheiro Barbosa, Patrick Anderson Santos, Larissa Alves Maciel, Patrício Lopes de Araujo Leite, Thiago Dos Santos Rosa, et al. "Is Lifelong Endurance Training Associated with Maintaining Levels of Testosterone, Interleukin-10, and Body Fat in Middle-Aged Males?" *Journal of Clinical and Translational Research* 7, no. 4 (2021): 450–55.

Herda, Ashley A., and Omid Nabavizadeh. "Short-Term Resistance Training in Older Adults Improves Muscle Quality: A Randomized Control Trial." *Experimental Gerontology* 145 (2021): 111195.

Hong, A. Ram, and Sang Wan Kim. "Effects of Resistance Exercise on Bone Health." *Endocrinology and Metabolism (Seoul)* 33, no. 4 (2018): 435–44.

Huxley, H. E. "Past, Present and Future Experiments on Muscle." *Philosophical Transactions: Biological Sciences* 355, no. 1396 (2020): 539–43.

Kamram, Ghazal. "Physical Benefits of (Salah) Prayer—Strengthen the Faith and Fitness." *Journal of Novel Physiotherapy and Rehabilitation* 2 (2018): 43–53.

Kemmler, W., A. Weissenfels, S. Willert, M. Shojaa, S. von Stengel, A. Filipovic, H. Kleinöder, J. Berger, and M. Fröhlich. "Efficacy and Safety of Low Frequency Whole-Body Electromyostimulation (WB-EMS) to Improve Health-Related Outcomes in Non-athletic Adults. A Systematic Review." *Frontiers in Physiology* 23, no. 9 (2018): 573.

Kemmler, Wolfgang, Heinz Kleinöder, and Michael Fröhlich. "Editorial: Whole-Body Electromyostimulation: A Training Technology to Improve Health and Performance in Humans?" *Frontiers in Physiology* 26, no. 11 (2020): 523.

Larsson, Lars, Hans Degens, Meishan Li, Leonardo Salviati, Young Il Lee, Wesley Thompson, James L. Kirkland, and Marco Sandri. "Sarcopenia: Aging-Related Loss of Muscle Mass and Function." *Physiology Review* 99, no. 1 (2019): 427–511.

Lazaro, R. J. "Effects of Lower Body Positive Pressure Treadmill Training on Balance, Mobility and Lower Extremity Strength of Community-Dwelling Older Adults: A Pilot Study." *Allied Health* 49, no. 2 (2020): e99–e103.

Lee, Se-Jin, Adam Lehar, Jessica U. Meir, Christina Koch, Andrew Morgan, Lara E. Warren, Renata Rydzik, et al. "Targeting Myostatin/

Activin A Protects Against Skeletal Muscle and Bone Loss during Spaceflight." *Proceedings of the National Academy of Sciences* 117, no. 38 (2020): 23942–51.

Lichtenberg, Theresa, Simon von Stengel, Cornel Sieber, and Wolfgang Kemmler. "The Favorable Effects of a High-Intensity Resistance Training on Sarcopenia in Older Community-Dwelling Men with Osteosarcopenia: The Randomized Controlled FrOST Study." *Clinical Interventions in Aging* 14 (2019): 2173–86.

Ludwig, Oliver, Joshua Berger, Torsten Schuh, Marco Backfisch, Stephan Becker, and Michael Fröhlich. "Can a Superimposed Whole-Body Electromyostimulation Intervention Enhance the Effects of a 10-Week Athletic Strength Training in Youth Elite Soccer Players?" *Journal of Sports Science and Medicine* 19, no. 3 (2020): 535–46.

May, A. K., A. P. Russell, and S. A. Warmington. "Lower Body Blood Flow Restriction Training May Induce Remote Muscle Strength Adaptations in an Active Unrestricted Arm." *European Journal of Applied Physiology* 118, no. 3 (2018): 617–27.

McKenna, C. F., A. F. Salvador, R. L. Hughes, S. E. Scaroni, R. A. Alamilla, A. T. Askow, S. A Paluska, et al. "Higher Protein Intake during Resistance Training Does Not Potentiate Strength, but Modulates Gut Microbiota, in Middle-Aged Adults: A Randomized Control Trial." *American Journal of Physiology, Endocrinology and Metabolism* 320, no. 5 (2021): E900–E913.

Mendonca, Goncalo, Carolina Vila-Chã, Carolina Teodósio, Andre D. Goncalves, Sandro R. Freitas, Pedro Mil-Homens, and Pedro Pezarat-Correia. "Contralateral Training Effects of Low-Intensity Blood-Flow Restricted and High-Intensity Unilateral Resistance Training." *European Journal of Applied Physiology* 121, no. 8 (2021): 2305–21.

Murphy, Caoileann C., Ellen M. Flanagan, Giuseppe De Vito, Davide Susta, Kathleen A. J. Mitchelson, Elena de Marco Castro, Joan M. G. Senden, et al. "Does Supplementation with Leucine-Enriched Protein Alone and in Combination with Fish-Oil-Derived N-3 Pufa Affect Muscle Mass, Strength, Physical Performance, and Muscle Protein Synthesis in Well-Nourished Older Adults? A Randomized, Double-Blind, Placebo-Controlled Trial." *American Journal of Clinical Nutrition* 113, no. 6 (2021): 1411–27.

National Aeronautics and Space Administration. "Muscle Atrophy." Accessed October 20, 2021. https://www.nasa.gov/pdf/64249main_ffs_factsheets_hbp_atrophy.pdf.

Nippard, Jeff. "How to Build Maximum Muscle (Explained in 5 Levels)." Accessed October 21, 2021. https://www.youtube.com/watch?v=lu_BObG6dj8.

Osama, Muhammad, and Reem Javed Malik. "Salat (Muslim Prayer) as a Therapeutic Exercise." *Journal of the Pakistan Medical Association* 69, no. 3 (2019): 399–404.

Park, Ji-Su, Sang-Hoon Lee, Sang-Hoon Jung, Jong-Bae Choi, and Young-Jin Jung. "Tongue Strengthening Exercise Is Effective in Improving the Oropharyngeal Muscles Associated with Swallowing in Community-Dwelling Older Adults in South Korea: A Randomized Trial." *Medicine (Baltimore)* 98, no. 40 (2019): e17304.

Peteiro, Jesus, and Alberto Bouzas-Mosquera. "Time to Climb 4 Flights of Stairs Provides Relevant Information on Exercise Testing Performance and Results." *Revista Española de Cardiologia (English Edition)* 74, no. 4 (2021): 354–55.

Rey-López, Juan Pablo, Emmanuel Stamatakis, Martin Mackey, Howard D. Sesso, and I-Min Lee. "Associations of Self-Reported Stair Climbing with All-Cause and Cardiovascular Mortality: The Harvard Alumni Health Study." *Preventative Medicine Reports* 15 (2019): 100938.

Schoenfeld, Brad J., Jozo Grgic, and James Krieger. "How Many Times Per Week Should a Muscle Be Trained to Maximize Muscle Hypertrophy? A Systematic Review and Meta-Analysis of Studies Examining the Effects of Resistance Training Frequency." *Journal of Sports Science* 37, no. 11 (2019): 1286–95.

Schoenfeld, B. J., and A. A. Aragon. "How Much Protein Can the Body Use in a Single Meal for Muscle-Building? Implications for Daily Protein Distribution." *Journal of the International Society of Sports Nutrition* 15 (2018): 10.

Smith, Tobie, Matthew Fedoruk, and Amy Eichner. "Performance-Enhancing Drug Use in Recreational Athletes." *American Family Physician* 103, no. 4 (2021): 203–4.

Stares, Aaron, and Mona Bains. "The Additive Effects of Creatine Supplementation and Exercise Training in an Aging Population: A Systematic Review of Randomized Controlled Trials." *Journal of Geriatric Physical Therapy* 43, no. 2 (2020): 99–112.

Stern, Marc. "The Fitness Movement and the Fitness Center Industry, 1960–2000." Accessed October 21, 2021. https://thebhc.org/sites/default/files/stern_0.pdf.

Stöllberger, Claudia, and Josef Finsterer. "Side Effects of and Contraindications for Whole-Body Electro-Myo-Stimulation: A Viewpoint." *BMJ Open Sport and Exercise Medicine* 5, no. 1 (2019): e000619.

Strasser, Barbara, Dominik Pesta, Jörn Rittweger, Johannes Burtscher, and Martin Burtscher. "Nutrition for Older Athletes: Focus on Sex-Differences." *Nutrients* 13, no. 5 (2021): 1409.

Tan, Jingwang, Xiaojian Shi, Jeremy Witchalls, Gordon Waddington, Allan C. Lun Fu, Sam Wu, Oren Tirosh, Xueping Wu, and Jia Han. "Effects of Pre-Exercise Acute Vibration Training on Symptoms of Exercise-Induced Muscle Damage: A Systematic Review and Meta-Analysis." *Journal of Strength and Conditioning Research* 36, no. 8 (August 2020): 2339–48.

Zehr, E. Paul. "The Man of Steel, Myostatin, and Super Strength." Accessed

October 21, 2021. https://blogs.scientificamerican.com/guest-blog/the-man-of-steel-myostatin-and-super-strength/.

Zoladz, Jerzy A. *Muscle and Exercise Physiology*. Cambridge, MA: Academic Press, 2018.

Chapter 7: Human Culture

Andersen, Jesper L., Peter Schjerling, and Bengt Saltin. "Muscle, Gene, and Athletic Performance." *Scientific American* 283, no. 3 (2000): 48–55.

Armstrong, Brock. "Do Amino Acids Build Bigger Muscles?" Accessed October 21, 2021. https://www.scientificamerican.com/article/do-amino-acids-build-bigger-muscles/.

Arnold, Carrie. "Virus Pumps Up Male Muscles—in Mice." Accessed October 20, 2021. https://www.scientificamerican.com/article/virus-pumps-up-male-muscles-in-mice/.

Beauty, Jivaka. "Easy Calf Reduction Methods for Beautiful Legs." Accessed October 21, 2021. https://beauty.jivaka.care/blogs/blog/slim-legs-with-calf-muscle-reduction-and-slimming-procedures.

Black, Ronald. "The Age of Sports—When Do Athletes Begin to Decline within Their Sport?" Accessed October 21, 2021. https://www.legitgamblingsites.com/blog/when-do-athletes-begin-to-decline-within-their-sport/.

Britannica. "Eugen Sandow." Accessed October 21, 2021. https://www.britannica.com/biography/Eugen-Sandow.

Brown-Séquard, C. E. "The Effects Produced on Man by Subcutaneous Injections of Liquid Obtained from the Testicles of Animals." *Lancet* 2 (1889): 105.

Brzeziańska, E., D. Domańska, and A. Jegier. "Gene Doping in Sport—Perspectives and Risks." *Biology of Sport* 31, no. 4 (2014): 251–59.

Carreyrou, John. "Flap over Doping Taints Another Group of Athletes—Pigeons." Accessed October 21, 2021. https://www.wsj.com/articles/SB110012337895470540.

Catlin, Oliver. "The WADA Prohibited List: A Guide and Temptation in Sports Nutrition." Accessed October 21, 2021. https://www.naturalproductsinsider.com/sports-nutrition/wada-prohibited-list-guide-and-temptation-sports-nutrition.

Cranswick, Ieuan. "Beyond the Muscles: Exploring the Meaning and Role of Muscularity in Identity." Accessed October 21, 2021. https://core.ac.uk/reader/222832880.

Dandoy, Christopher, and Rani S. Gereige. "Performance-Enhancing Drugs." *Pediatrics in Review* 33, no. 6 (2012): 265–72.

Davis, Josh. "Eugen Sandow: A Body Worth Immortalising." Accessed October 21, 2021. https://www.nhm.ac.uk/discover/eugen-sandow-a-body-worth-immortalising.html.

Economist. "Sport Is Still Rife with Doping." Accessed October 21, 2021.

https://www.economist.com/science-and-technology/2021/07/14/sport-is-still-rife-with-doping.

Giaimo, Cara. "Were Colonial Men Obsessed with Their Calves?" Accessed October 21, 2021. https://www.atlasobscura.com/articles/colonial-calves-men-fashion-myth.

Gill, Michael. "A History of Stone Lifting and Strongman." Accessed October 21, 2021. https://barbend.com/strongman-stone-history/.

Government of Canada. "What Happens to Muscles in Space." Accessed October 20, 2021. https://www.asc-csa.gc.ca/eng/sciences/osm/muscles.asp.

Greenemeier, Larry. "Unnatural Selection: Muscles, Genes and Genetic Cheats." Accessed October 20, 2021. https://www.scientificamerican.com/article/muscles-genes-cheats-2012-olympics-london/.

Huebner, Marianne, and Aris Perperoglou. "Performance Development from Youth to Senior and Age of Peak Performance in Olympic Weightlifting." *Frontiers in Physiology* 10 (2019): 1121.

Longman, Jeré. "85-Year-Old Marathoner Is So Fast That Even Scientists Marvel." Accessed October 21, 2021. https://www.nytimes.com/2016/12/28/sports/ed-whitlock-marathon-running.html.

Malcata, Rita M., and Will G. Hopkins. "Variability of Competitive Performance of Elite Athletes: A Systematic Review." *Sports Medicine* 44, no. 12 (2014): 1763–74.

Malta, Elvis S., Yago M. Dutra, James R. Broatch, David J. Bishop, and Alessandro M. Zagatto. "The Effects of Regular Cold-Water Immersion Use on Training-Induced Changes in Strength and Endurance Performance: A Systematic Review with Meta-Analysis." *Sports Medicine* 1, no. 1 (2021): 161–74.

Melville, Herman. *Typee*. New York: Wiley and Putnam, 1846.

Momaya, Amit, Marc Fawal, and Reed Estes. "Performance-Enhancing Substances in Sports: A Review of the Literature." *Sports Medicine* 45, no. 4 (2015): 517–31.

Morgan, Chance. "The History of Strength Training." Accessed October 21, 2021. http://thesportdigest.com/archive/article/history-strength-training.

Murden, Sarah. "Bums, Tums and Downy Calves." Accessed October 21, 2021. https://georgianera.wordpress.com/2014/07/22/bums-tums-and-downy-calves/.

Petersen, A. C., and J. J. Fyfe. "Post-Exercise Cold Water Immersion Effects on Physiological Adaptations to Resistance Training and the Underlying Mechanisms in Skeletal Muscle: A Narrative Review." *Frontiers in Sports and Active Living* 8, no. 3 (2021): 660291.

Robinson, Joshua. "How to Exhaust a Tour de France Racer: Ask Him to Take a Walk." Accessed October 20, 2021. https://www.wsj.com/articles/tour-de-france-cycle-racers-walk-10–000-steps-11600275036?mod=searchresults_pos15&page=2.

Schwarzenegger, A. *The New Encyclopedia of Modern Bodybuilding*. New York: Fireside/Simon and Schuster, 1999.

Segre, Paolo S., Jean Potvin, David E. Cade, John Calambokidis, Jacopo Di Clemente, Frank E. Fish, Ari S. Friedlaender, William T. Gough, et al. "Energetic and Physical Limitations on the Breaching Performance of Large Whales." *eLife* 9 (2020): e51760.

Stone, Ken. "Dash of History: 100-Year-Old Sets 5 World Records." Accessed October 21, 2021. https://timesofsandiego.com/sports/2015/09/20/dash-of-history-100-year-old-sets-5-world-records/.

Tan, Jingwang, Xiaojian Shi, Jeremy Witchalls, Gordon Waddington, Allan C. Lun Fu, Sam Wu, Oren Tirosh, Xueping Wu, and Jia Han. "Effects of Pre-Exercise Acute Vibration Training on Symptoms of Exercise-Induced Muscle Damage: A Systematic Review and Meta-Analysis." *Journal of Strength and Conditioning Research* 36 (2020): 2339–48.

Twardziak, Kelly. "10 Facts About Bodybuilding Legend Eugen Sandow." Accessed October 21, 2021. https://www.muscleandfitness.com/athletes-celebrities/news/10-facts-about-bodybuilding-legend-eugen-sandow/.

Vance, John. "Effective Drug Policies for Racing Pigeons." Accessed October 21, 2021. https://www.pigeonracingpigeon.com/pigeon-racing/effective-drug-policies-for-racing-pigeons/.

Wise, J. *Extreme Fear: The Science of Your Mind in Danger*. New York: Palgrave Macmillan, 2011.

World Anti-Doping Agency. "World Anti-Doping Code International Standard Prohibited List 2021." Accessed October 21, 2021. https://www.wada-ama.org/sites/default/files/resources/files/2021list_en.pdf.

Yeager, Selene. "This Woman Just Biked at 184 MPH to Smash the Bicycle Speed Record." Accessed October 20, 2021. https://www.bicycling.com/news/a23281242/denise-mueller-korenek-breaks-bicycle-speed-record/.

Chapter 8: Discomforts and Diseases

Afonso, José, Filipe Manuel Clemente, Fábio Yuzo Nakamura, Pedro Morouço, Hugo Sarmento, Richard A. Inman, and Rodrigo Ramirez-Campillo. "The Effectiveness of Post-Exercise Stretching in Short-Term and Delayed Recovery of Strength, Range of Motion and Delayed Onset Muscle Soreness: A Systematic Review and Meta-Analysis of Randomized Controlled Trials." *Frontiers in Physiology* 12 (2021): 677581.

Agergaard, J., S. Leth, T. H. Pedersen, T. Harbo, J. U. Blicher, P. Karlsson, L. Østergaard, H. Andersen, and H. Tankisi. "Myopathic Changes in Patients with Long-Term Fatigue after COVID-19." *Clinical Neurophysiology* 132, no. 8 (2021): 1974–81.

Bunnell, Sterling. "Restoring Flexion to the Paralytic Elbow." *Journal of Bone and Joint Surgery* 33, no. 3 (1951): 569.

Choi, Ji Yun, Hyo Joon Kim, and Seong Yong Moon. "Management of the Paralyzed Face Using Temporalis Tendon Transfer via Intraoral and

Transcutaneous Approach." *Maxillofacial Plastic and Reconstructive Surgery* 40, no. 1 (2018): 24.

Fang, Wang, and Yasaman Nasir. "The Effect of Curcumin Supplementation on Recovery Following Exercise-Induced Muscle Damage and Delayed-Onset Muscle Soreness: A Systematic Review and Meta-Analysis of Randomized Controlled Trials." *Physiotherapy Research* 35, no. 4 (2021): 1768–81.

Fernández-Lázaro, D., J. Mielgo-Ayuso, J. Seco Calvo, A. Córdova Martínez, A. Caballero García, and C. I. Fernandez-Lazaro. "Modulation of Exercise-Induced Muscle Damage, Inflammation, and Oxidative Markers by Curcumin Supplementation in a Physically Active Population: A Systematic Review." *Nutrients* 12, no. 2 (2020): 501.

Gawecki, Maciej. "Adjustable Versus Nonadjustable Sutures in Strabismus Surgery—Who Benefits the Most?" *Journal of Clinical Medicine* 9, no. 2 (2020): 292–305.

Giuriato, G., A. Pedrinolla, F. Schena, and M. Venturelli. "Muscle Cramps: A Comparison of the Two-Leading Hypothesis." *Journal of Electromyography and Kinesiology* 41 (2018): 89–95.

Heiss, Rafael, Christoph Lutter, Jürgen Freiwald, Matthias W. Hoppe, Casper Grim, Klaus Poettgen, Raimund Forst, et al. "Advances in Delayed-Onset Muscle Soreness (DOMS)—Part II: Treatment and Prevention." *Sportsverletzung Sportschaden* 33, no. 1 (2019): 21–29.

Kolata, Gina. "A Very Muscular Baby Offers Hope Against Diseases." Accessed October 20, 2021. https://www.nytimes.com/2004/06/24/us/a-very-muscular-baby-offers-hope-against-diseases.html.

Ma, Fenghao, Yingqi Li, Jinchao Yang, Xidian Li, Na Zeng, and RobRoy L. Martin. "The Effectiveness of Low Intensity Exercise and Blood Flow Restriction without Exercise on Exercise Induced Muscle Damage: A Systematic Review." *Physical Therapy in Sport* 46 (2020): 77–88.

Maughan, R. J., and S. M. Shirreffs. "Muscle Cramping during Exercise: Causes, Solutions, and Questions Remaining." *Sports Medicine* 49, Supplement 2 (2019): 115–24.

Mueller, Amber L., Andrea O'Neill, Takako I. Jones, Anna Llach, Luis Alejandro Rojas, Paraskevi Sakellariou, Guido Stadler, et al. "Muscle Xenografts Reproduce Key Molecular Features of Facioscapulohumeral Muscular Dystrophy." *Experimental Neurology* 320 (2019): 113011.

Négyesi, János, Li Yin Zhang, Rui Nian Jin, Tibor Hortobágyi, and Ryoichi Nagatomi. "A Below-Knee Compression Garment Reduces Fatigue-Induced Strength Loss but Not Knee Joint Position Sense Errors." *European Journal of Applied Physiology* 121, no. 1 (2021): 219–29.

Okabe, Yuka Tsukagoshi, Shinobu Shimizu, Yukihiro Suetake, Hisako Matsui-Hirai, Shizuka Hasegawa, Keisuke Takanari, Kazuhiro Toriyama, et al. "Biological Characterization of Adipose-Derived Regenerative Cells Used for the Treatment of Stress Urinary Incontinence." *International Journal of Urology* 28, no. 1 (2021): 115–24.

Puntillo, Filomena, Mariateresa Giglio, Antonella Paladini, Gaetano Perchiazzi, Omar Viswanath, Ivan Urits, Carlo Sabbà, et al. "Pathophysiology of Musculoskeletal Pain: A Narrative Review." *Therapeutic Advances in Musculoskeletal Disease* 12 (2021): 1759720X21995067.

Romero-Parra, Nuria, Rocío Cupeiro, Victor M. Alfaro-Magallanes, Beatriz Rael, Jacobo Á. Rubio-Arias, Ana B. Peinado, Pedro J. Benito, and IronFEMME Study Group. "Exercise-Induced Muscle Damage during the Menstrual Cycle: A Systematic Review and Meta-Analysis." *Journal of Strength and Conditioning Research* 35, no. 2 (2021): 549–61.

Roth, Stephen M. "Why Does Lactic Acid Build Up in Muscles? And Why Does It Cause Soreness?" Accessed October 20, 2021. https://www.scientificamerican.com/article/why-does-lactic-acid-buil/.

Rutecki, G. W., A. J. Ognibene, and J. D. Geib. "Rhabdomyolysis in Antiquity. From Ancient Descriptions to Scientific Explication." *Pharos* 61, no. 2 (1998): 18–22.

Saber, Mohamed. "Myositis Ossificans." Accessed October 21, 2021. https://radiopaedia.org/articles/myositis-ossificans-1?lang=us.

Selva-O'Callaghan, Albert, Marcelo Alvarado-Cardenas, Iago Pinal-Fernández, Ernesto Trallero-Araguás, José Cesar Milisenda, María Ángeles Martínez, Ana Marín, Moisés Labrador-Horrillo, et al. "Statin-Induced Myalgia and Myositis: An Update on Pathogenesis and Clinical Recommendations." *Expert Review of Clinical Immunology* 14, no. 3 (2018): 215–24.

Stefanelli, Lucas, Evan J. Lockyer, Brandon W. Collins, Nicholas J. Snow, Julie Crocker, Christopher Kent, Kevin E. Power, et al. "Delayed-Onset Muscle Soreness and Topical Analgesic Alter Corticospinal Excitability of the Biceps Brachii." *Medicine and Science in Sports and Exercise* 51, no. 11 (2019): 2344–56.

Swash M., D. Czesnik, and M. de Carvalho. "Muscle Cramp: Causes and Management." *European Journal of Neurology* 26, no. 2 (2019): 214–21.

Thilo, Jürgen Freiwald, Matthias Wilhelm Hoppe, Christoph Lutter, Raimund Forst, Casper Grim, Wilhelm Bloch, Moritz Hüttel, et al. "Advances in Delayed-Onset Muscle Soreness (DOMS): Part I: Pathogenesis and Diagnostics." *Sportsverletzung Sportschaden* 32, no. 4 (2019): 243–50.

Walton, Zeke, Milton Armstrong, Sophia Traven, and Lee Leddy. "Pedicled Rotational Medial and Lateral Gastrocnemius Flaps: Surgical Technique." *Journal of the American Academy of Orthopaedic Surgeons* 25, no. 11 (2017): 744–51.

Wan, Jing-jing, Zhen Qin, Peng-yuan Wang, Yang Sun, and Xia Liu. "Muscle Fatigue: General Understanding and Treatment." *Experimental and Molecular Medicine* 49, no. 10 (2017): e384.

Ward, Natalie C., Gerald F. Watts, and Robert H. Eckel. "Statin Toxicity Mechanistic Insights and Clinical Implications." *Circulation Research* 124 (2019): 328–50.

Chapter 9: Zoological Survey

Boas, J. E. V., and Simon Paulli. *The Elephant's Head; Studies in the Comparative Anatomy of the Organs of the Head of the Indian Elephant and Other Mammals.* Copenhagen: Carlsberg Fund, 1908–1925.

Boswall, Jeffery. "How Birds Sing." Accessed October 21, 2021. https://www.bl.uk/the-language-of-birds/articles/how-birds-sing.

Britannica. "Syrinx Bird Anatomy." Accessed October 20, 2021. https://www.britannica.com/science/syrinx-bird-anatomy.

Burrows, Malcolm. "Jumping Performance of Froghopper Insects." *Journal of Experimental Biology* 209, no. 23 (2006): 4607–21.

Callier, Viviane. "Too Small for Big Muscles, Tiny Animals Use Springs." Accessed October 18, 2021. https://www.scientificamerican.com/article/too-small-for-big-muscles-tiny-animals-use-springs/.

Chantler, P. D. "Scallop Adductor Muscles: Structure and Function." *Developments in Aquaculture and Fisheries Science* 35 (2006): 229–316.

Chatfield, Matthew. "Woodpeckers' Tongues Fit the Bill." Accessed October 21, 2021. https://naturenet.net/blogs/2008/02/10/woodpeckers-tongues-fit-the-bill/.

Chen, Natalie. "A Flexible Body Allows the Earthworm to Burrow Through Soil." Accessed March 1, 2022. https://asknature.org/strategy/a-flexible-body-allows-the-earthworm-to-burrow-through-soil/.

Dewhurst, H. W. *The Natural History of the Order Cetacea, and the Oceanic Inhabitants of the Arctic Regions.* London: H. W. Dewhurst, 1834.

Gil, Kelsey N., Margo A. Lillie, A. Wayne Vogl, and Robert E. Shadwick. "Rorqual Whale Nasal Plugs: Protecting the Respiratory Tract against Water Entry and Barotrauma." *Journal of Experimental Biology* 223, no. 4 (2020): jeb219691.

Gill, Robert E., Jr., T. Lee Tibbitts, David C. Douglas, Colleen M. Handel, Daniel M. Mulcahy, Jon C. Gottschalk, Nils Warnock, et al. "Extreme Endurance Flights by Landbirds Crossing the Pacific Ocean: Ecological Corridor Rather Than Barrier?" *Proceedings: Biological Sciences* 276, no. 1656 (2009): 447–57.

Guglielmo, Christopher G., Theunis Piersma, and Tony D. Williams. "A Sport-Physiological Perspective on Bird Migration: Evidence for Flight-Induced Muscle Damage." *Journal of Experimental Biology* 204, no. 15 (2001): 2683–90.

Harrison, Robert. "On the Anatomy of the Elephant." *Proceedings of the Royal Irish Academy* 3 (1844–1847): 392–98.

Higham, Timothy E., and Anthony P. Russell. "Flip, Flop and Fly: Modulated Motor Control and Highly Variable Movement Patterns of Autotomized Gecko Tails." *Biology Letters* 6, no. 1 (2010): 70–73.

Kaminski, Juliane, Bridget M. Waller, Rui Diogo, Adam Hartstone-Rose, and Anne M. Burrows. "Evolution of Facial Muscle Anatomy in Dogs."

Proceedings of the National Academy of Sciences 116, no. 29 (2019): 14677–81.

Kier, William M. "The Musculature of Coleoid Cephalopod Arms and Tentacles." *Frontiers in Cell and Developmental Biology* 18, no. 4 (2016): 10.

Landys-Cianelli, M. M., T. Piersma, and J. Jukema. "Strategic Size Changes of Internal Organs and Muscle Tissue in the Bar-Tailed Godwit during Fat Storage on a Spring Stopover Site." *Functional Ecology* 17, no. 2 (2003): 151–59.

Langworthy, Orthello R. "A Morphological Study of the Panniculus Carnosus and Its Genetical Relationship to the Pectoral Musculature in Rodents." *American Journal of Anatomy* 35, no. 2 (1925): 283–302.

Naldaiz-Gastesi, Neia, Ola A. Bahri, Adolfo López de Munain, Karl J. A. McCullagh, and Ander Izeta. "The Panniculus Carnosus Muscle: An Evolutionary Enigma at the Intersection of Distinct Research Fields." *Journal of Anatomy* 233, no. 3 (2018): 275–88.

Patel, Amir, Edward Boje, Callen Fisher, Leeann Louis, and Emily Lane. "Quasi-Steady State Aerodynamics of the Cheetah Tail." *Biology Open* 5, no. 8 (2016): 1072–76.

Schmidt, Marc F., and J. Martin Wild. "The Respiratory-Vocal System of Songbirds: Anatomy, Physiology, and Neural Control." *Progress in Brain Research* 212 (2014): 297–335.

Tramacere, F., L. Beccai, M. Kuba, A. Gozzi, A. Bifone, and B. Mazzolai. "The Morphology and Adhesion Mechanism of *Octopus Vulgaris* Suckers." *PLOS One* 8, no. 6 (2013): 65074.

Wilson, J. F., U. Mahajan, S. A. Wainwright, and L. J. Croner. "A Continuum Model of Elephant Trunks." *Journal of Biomechanical Engineering* 113, no. 1 (1991): 79–84.

Chapter 10: Other Force Producers

Abraham, Zachary, Emma Hawley, Daniel Hayosh, Victoria A. Webster-Wood, and Ozan Akkus. "Kinesin and Dynein Mechanics: Measurement Methods and Research Applications." *Journal of Biomechanical Engineering* 140, no. 2 (2018): 0208051–02080511.

Berg, Howard C. "The Rotary Motor of Bacterial Flagella." *Annual Review of Biochemistry* 72 (2003): 19–54.

Botanical Society of America. "The Mysterious Venus Flytrap." Accessed October 21, 2021. https://www.wsj.com/articles/olympic-runner-failed-drug-test-burrito-shelby-houlihan-11623867500?mod=hp_major_pos1#cxrecs_s.

Chen, Xiang-Jun, Huan Xu, Helen M. Cooper, and Yaobo Liu. "Cytoplasmic Dynein: A Key Player in Neurodegenerative and Neurodevelopmental Diseases." *Science China Life Sciences* 57, no. 4 (2014): 372–77.

Cooper, G. M., and M. A. Sunderland. *The Cell: A Molecular Approach.* 2nd ed. Washington, DC: American Society of Microbiology, 2000.

Darwin, Charles, and Francis Darwin. *The Power of Movement in Plants.* New York: D. Appleton, 1898.

Duan, Zhongrui, and Motoki Tominaga. "Actin-Myosin XI: An Intracellular Control Network in Plants." *Biochemical and Biophysical Research Communications* 506, no. 2 (2018): 403–8.

Ebrahimkhani, Mo R., and Michael Levin. "Synthetic Living Machines: A New Window on Life." *iScience* 24, no. 5 (2021): 102502.

Ellis, C. H. "The Mechanism of Extension of the Legs of Spiders." *Biological Bulletin* 86 (1944): 41–50.

Ghoshdastider, U., S. Jiang, D. Popp, and R. C. Robinson. "In Search of the Primordial Actin Filament." *Proceedings of the National Academy of Sciences* 112, no. 30 (2015): 9150–51.

Gubert, Carolina, and Anthony J. Hannan. "Exercise Mimetics: Harnessing the Therapeutic Effects of Physical Activity." *Nature Reviews Drug Discovery* 20, no. 11 (2021): 862–79.

Gunning, P. W., U. Ghoshdastider, S. Whitaker, D. Popp, and R. C. Robinson. "The Evolution of Compositionally and Functionally Distinct Actin Filaments." *Journal of Cell Science* 128, no. 11 (2015): 2009–19.

Hagihara, Takuma, and Masatsugu Toyota. "Mechanical Signaling in the Sensitive Plant *Mimosa pudica* L." *Plants (Basel)* 9, no. 5 (2020): 587.

Hartman, M. Amanda, and James A. Spudich. "The Myosin Superfamily at a Glance." *Journal of Cell Science* 125, no. 7 (2012): 1627–32.

Hawley, John A., Michael J. Joyner, and Daniel J. Green. "Mimicking Exercise: What Matters Most and Where to Next?" *Journal of Physiology* 599, no. 3 (2021): 791–802.

Hendriks, Adam G. "Low Efficiency Spotted in a Molecular Motor." *Physics* 11 (2018): 120.

Hinz, Boris, and David Lagares. "Evasion of Apoptosis by Myofibroblasts: A Hallmark of Fibrotic Diseases." *Nature Reviews Rheumatology* 16 (2020): 11–31.

Hoffmeister, Dirk, and Markus Gressler. *Biology of the Fungal Cell.* New York: Springer, 2019.

Iino, Ryota, Kazushi Kinbara, and Zev Bryant. "Introduction: Molecular Motors." *Chemical Reviews* 120, no. 1 (2020): 1–4.

Krakhmal, N. V., M. V. Zavyalova, E. V. Denisov, S. V. Vtorushin, and V. M. Perelmuter. "Cancer Invasion: Patterns and Mechanisms." *Acta Naturae* 7, no. 2 (2015): 17–28.

Kuek, Li Eon, and Robert J. Lee. "First Contact: The Role of Respiratory Cilia in Host-Pathogen Interactions in the Airways." *American Journal of Physiology—Lung, Cell, and Molecular Physiology* 319, no. 4 (2020): L603–L619.

Kuiken, Todd A., Ann K. Barlow, Levi Hargrove, and Gregory A. Dumanian. "Targeted Muscle Reinnervation for the Upper and Lower Extremity." *Techniques in Orthopaedics* 32, no. 2 (2017): 109–16.

Kurth, Elizabeth G., Valera V. Peremyslov, Hannah L. Turner, Kira S.

Makarova, Jaime Iranzo, Sergei L. Mekhedov, Eugene V. Koonin, et al. "Myosin-Driven Transport Network in Plants." *Proceedings of the National Academy of Sciences* 114, no. 8 (2017): E1385–E1394.

La Porta, Caterina, and Stefano Zapperi. *Cell Migrations: Causes and Functions*. New York: Springer, 2019.

Lauga, Eric. *The Fluid Dynamics of Cell Motility*. Cambridge: Cambridge University Press, 2020.

Lindås, Ann-Christin, Karin Valegård, and Thijs J. G. Ettema. "Archaeal Actin-Family Filament Systems." *Subcellular Biochemistry* 84 (2017): 379–92.

Lou, Sunny S., Andrew S. Kennard, Elena F. Koslover, Edgar Gutierrez, Alexander Groisman, and Julie A. Theriot. "Elastic Wrinkling of Keratocyte Lamellipodia Driven by Myosin-Induced Contractile Stress." *Biophysical Journal* 120, no. 9 (2021): 1578–91.

McGrath, Jamis, Roy Pallabi, and Benjamin J. Perrin. "Stereocilia Morphogenesis and Maintenance Through Regulation of Actin Stability." *Seminars in Cell and Developmental Biology* 65 (2017): 88–95.

Morell, Maria, A. Wayne Vogl, Lonneke L. IJsseldijk, Marina Piscitelli-Doshkov, Ling Tong, Sonja Ostertag, Marissa Ferriera, et al. "Echolocating Whales and Bats Express the Motor Protein Prestin in the Inner Ear: A Potential Marker for Hearing Loss." *Frontiers in Veterinary Science* 7 (2020): 429.

Mueller, Sabine. "Plant Cell Division—Defining and Finding the Sweet Spot for Cell Plate Insertion." *Current Opinion in Cell Biology* 60 (2019): 9–18.

Nangole, Ferdinand, and George W. Agak. "Keloid Pathology: Fibroblast or Inflammatory Disorder?" *JPRAS Open* 22 (2019): 44–54.

Okimura, Chika, Atsushi Taniguchi, Shigenori Nonaka, and Yoshiaki Iwadate. "Rotation of Stress Fibers as a Single Wheel in Migrating Fish Keratocytes." *Scientific Reports* 2018, no. 8 (2018): 10615.

Parry, Wynne. "How the Venus Flytrap Kills and Digests Its Prey." Accessed October 21, 2021. https://www.livescience.com/15910-venus-flytrap-carnivorous.html.

Poppinga, Simon, Carmen Weisskopf, Anna Sophia Westermeier, Tom Masselter, and Thomas Speck. "Fastest Predators in the Plant Kingdom: Functional Morphology and Biomechanics of Suction Traps Found in the Largest Genus of Carnivorous Plants." *AoB Plants* 8 (2015): plv140.

Ryan, Jennifer M., and Andreas Nebenführ. "Update on Myosin Motors: Molecular Mechanisms and Physiological Functions." *Plant Physiology* 176, no. 1 (2018): 119–27.

Sahi, Vaidurya Pratap, and František Baluška. *Concepts in Cell Biology—History and Evolution*. New York: Springer, 2018.

Trepat, Xavier, Zaozao Chen, and Ken Jacobson. "Cell Migration." *Comprehensive Physiology* 2, no. 4 (2012): 2369–92.

Wang, Yifeng, and Hua Li. "Bio-Chemo-Electro-Mechanical Modelling of the Rapid Movement of *Mimosa pudica*." *Bioelectrochemistry* 134 (2020): 107533.

Wayne, Randy O. "Actin- and Microfilament-Mediated Processes." In *Plant Cell Biology: From Astronomy to Zoology*, 2nd ed. Cambridge, MA: Academic Press, 2018.

Woodhouse, Francis G., and Raymond E. Goldstein. "Cytoplasmic Streaming in Plant Cells Emerges Naturally by Microfilament Self-Organization." *Proceedings of the National Academy of Sciences* 110, no. 35 (2013): 14132–37.

Wooley, David M. "Flagellar Oscillation: A Commentary on Proposed Mechanisms." *Biologic Reviews of the Cambridge Philosophical Society* 85, no. 3 (2010): 453–70.

Yamada, Kenneth M., and Michael Sixt. "Mechanisms of 3D Cell Migration." *Nature Reviews of Molecular and Cell Biology* 20, no. 12 (2019): 738–52.

Yamaguchi, Takami, Takuji Isikawa, and Yohsuke Imai. *Integrated Nano-Biomechanics*. Amsterdam: Elsevier, 2018.

Zheng, Huang, Yuxin Liu, and Zi Chen. "Fast Motion of Plants: From Biomechanics to Biomimetics." *Journal of Postdoctoral Research* 1, no. 2 (2013): 40–50.

Illustration Credits

7: J. G. De Lint. *Atlas of the History of Medicine*, Vol. 1: *Anatomy*. New York: Hoeber, 1926.

12, right: Visible Human Male Project. Courtesy of the National Library of Medicine.

14: Courtesy of the National Library of Medicine.

15: Visible Human Male Project. Courtesy of the National Library of Medicine.

80: Photomicrograph courtesy of Scott D. Nelson, MD.

107: Courtesy of Benjamin Plotkin, MD.

125: Shutterstock.

132, left: Alex Feldstein, Creative Commons.

134, bottom: Courtesy of the Dallas Museum of Art.

147, lower images: Courtesy of Paul N. Chugay, MD.

166: Joh-co, Wikipedia Creative Commons 3.0.

174: Sterling Bunnell. "Restoring Flexion to the Paralytic Elbow." *Journal of Bone and Joint Surgery* 33, no. 3 (July 1951): 569.

175: Subject granted permission; photos courtesy of Roger L. Simpson, MD.

177: Courtesy of the National Cancer Institute.

199, left: Htirgan, Creative Commons 3.

206, C: David M. Wooley. "Flagellar Oscillation: A Commentary on Proposed Mechanisms." *Biologic Reviews of the Cambridge Philosophical Society* 85, no. 3 (2010): 453–70.

209: Shutterstock.

217: Schokraie E, Warnken U, Hotz-Wagenblatt A, Grohme MA, Hengherr S, et al., Creative Commons.

Index